Instructor's Manual: ANSWER GUIDE

for

Biology in the Laboratory
Third Edition

Doris R. Helms
Carl W. Helms
Barbara J. Speziale

Clemson University

W. H. Freeman and Company
New York

SUPPLEMENTS EDITOR: *Patrick Shriner*
ASSOCIATE EDITOR: *Debra Siegel*
ADMINISTRATIVE ASSISTANT: *Ceserina Pugliese*
COVER DESIGN: *Vicki Tomaselli*
PRODUCTION COORDINATOR: *Paul W. Rohloff*

ISBN: 0-7167-3236-X

Printed in the United States of America

First printing 1998

TABLE OF CONTENTS

FOREWORD

There are no absolute answers in science—only observations, data, and interpretations that support or negate a hypothesis. Nevertheless, we are pleased to offer an "answer" guide as a supplement to *Biology in the Laboratory, 3rd ed.*

Within each laboratory, answers are arranged by exercise and by page number. Using the *Answer Guide* together with the *Preparator's Guide* will allow you and your students to address any questions you may have about experimental procedures and outcomes. Both guides should also help you identify the origins of students' difficulties and errors; ensure that the work of your laboratory assistants is efficient, effective, and consistent; and assist you with grading.

The sample data, graphs, tables, and quantitative data analyses included here were derived directly from classroom experiments conducted by students. Guidelines are provided for all questions in *Biology in the Laboratory, 3rd ed.*, many of which are designed to foster critical thinking and the interpretation of data. Other questions extend laboratory topics and require students to assemble an answer from other resources. The examples of hypotheses and experimental procedures offered throughout the laboratory exercises and for **Extending Your Investigation** activities only hint at the diversity of potential valid ideas and approaches.

Answers for the BioBytes simulation exercises, *Alien, Cycle, Dueling Alleles,* and *Seedling,* are also included at the end of this *Answer Guide,* followed by answers provided to problems and questions contained in the Appendices.

We have designed *Biology in the Laboratory, 3rd ed.* and its *Answer Guide* and *Preparator's Guide* to provide opportunity, encouragement, and support for the exploration and enjoyment of biology. Please let us know how well we've succeeded and how we can improve. If you have a particularly good set of student data or preparation short-cuts or alternatives, please share them with us and others by sending it to us by e-mail so we can include it on our developing Web Homepage, http://www/whfreeman.com/biolab/helms. Check this site for additional information and corrections to *Biology in the Laboratory, 3rd ed.* and the supporting *Instructor's Manuals.*

Doris R. Helms, Biology Program, 330 Long Hall, BIOL110@clemson.edu, 864-656-2418 (voice), 864-656-3839 (FAX)

Carl W. Helms, Biological Sciences, 132 Long Hall, cwhelms@clemson.edu, 864-656-4224

Biology Program, 330 Long Hall, Clemson University, Clemson, SC 29634

December, 1997

Instructor's Manual:
ANSWER GUIDE

Laboratory I

Science—A Process

EXERCISE A   The Scientific Method

p. I-4   1. Step 1. Observations—"I got 70% on the first biology exam. I want a better grade than that. What am I doing wrong? What can I do to improve my grade?"

Step 2. Hypothesis—"I don't know the material well enough because I'm not reviewing my notes after every lecture and lab."

Step 3. Predictions—"My grade will improve if I start reviewing my notes after every lecture and lab."

Step 4. Independent and dependent variables—"The independent variable is my study routine. The dependent variable is my biology grade on the exams."

Step 5. Experimental treatments—"One treatment will be my study routine and results from the first exam. The second treatment will be my changed study routine and, hopefully, higher grades on the second exam."

Step 6. Experimental design—"To prepare for the second exam, I will do everything the same way as I did for the first exam, with the exception that I will review my notes after every lecture and laboratory."

Step 7. Results—"I did the readings, went to class, took notes (just as I had for the first exam), *and* I reviewed my notes after every lecture and lab. Then I took the exam and compared my grade with my grade on the first exam."

Step 8. Analysis—"I got an 80%--much better than the 70% on the first exam!"

Step 9. Refinements—"I think I could do better than an 80% if I took notes on the reading assignments instead of just reading them. I return to step 2."

2. "If my grade does not improve, then failure to review my notes after each lecture and lab is not a problem. Perhaps I need more extensive review, or perhaps the solution lies in some other direction."

EXERCISE B   Reaction Time Experiments: Making Observations

p. I-4

Stimulus	Observed Response
car in front of me slows down	I step on the brake
I hear my alarm clock go off	I turn the alarm off and get up
cold winds chill me	I pull my coat around myself
I feel a tap on the shoulder	I turn
I feel thirst	I drink some water

PART 1   Making Observations (*for Laboratories Using Computers*)

p. I-5   7. Results—some possible observations are:

I'm a lot slower on low-contrast spot-the-dot than on high-contrast.

The more letters on the list in symbol recognition, the slower my response and the more errors I make.

I seem to be faster on the auditory stimuli than on the visual ones.

If there's background noise, my reaction is slower on auditory stimuli.

Conversation in the background does not affect my reaction time to simple stimuli like "X at a known location," but really distracts me when I'm using complex stimuli like symbol recognition.

People who wear glasses seem to have slower reaction times than people who don't wear glasses.

People who didn't get much sleep the night before do poorly on complex stimuli like symbol recognition

PART 2   Making Observations (*for Laboratories Not Using Computers*)

p. I-6   3. Results—some possible observations are:

If someone is distracted at the moment the ruler is dropped, the reaction time is very poor.

People with small hands tend to do better than people with big hands.

Sometimes the ruler slides down further even after the fingers have closed on it, so finger strength rather than just reaction time may be involved.

EXERCISE C   The Reaction Time Experiments: Formulating a Hypothesis

PART 1   Formulating a Hypothesis (*for Laboratories Using Computers*)

Example 1—Paired Treatments

p. I-7   Stimulus used: high-contrast spot-the-dot.

Observation: the dominant hand appears to be faster.

Hypothesis: reaction times for the dominant and non-dominant hand will be different.

Null hypothesis: reaction times for the dominant and non-dominant time will not be different.

Example 2—Unpaired Treatments

p. I-7   Stimulus used: symbol recognition (5 letters).

Observation: piano players seem to be more dexterous.

Hypothesis: piano players will have a faster reaction time than non-piano players

Null hypothesis: reaction times for piano players and non-piano players will not be different.

PART 2   Formulating a Hypothesis (*for Laboratories Not Using Computers*)

p. I-8   Stimulus used: dropping the ruler.

Observation: the dominant hand appears to be faster.

Hypothesis: reaction times for the dominant and non-dominant hand will be different.

Null hypothesis: reaction times for the dominant and non-dominant hand will not be different.

EXERCISE D   The Reaction Time Experiments: Developing an Experimental Design

PART 1   Identifying Variables (*for all Laboratories, Using or Not Using Computers*)

p. I-8   a. Time.

---

p. I-9 **b.** Reaction time (or another variable being related to time of day).

c. Other variables could influence the results by introducing bias.

d. Nutrition, disturbance, lighting, experimental set-up—essentially anything that could introduce bias or confound the results obtained.

Dependent variable(s): reaction time.

Independent variable: hand use.

Controlled (standardized) variables: stimulus, distance from keyboard and screen, same lighting, absence of distractions, etc.

PART 2 Defining Experimental Treatments (*for All Laboratories, Using or Not Using Computers*)

Example 1—Paired Treatments

p. I-10 Treatment 1: All the individuals in a lab group, each using his or her dominant hand to reach for the stimulus.

Treatment 2: The same individuals, but this time each person uses his or her non-dominant hand.

a. No.

b. You are comparing results from two types of hand-dominance (independent variable). There is no control group. Both groups are treatments. Although you might think that the non-dominant hand could be used as a control since reaction times will vary from individual to individual, this should be used as a treatment group and not as a control.

Example 2—Unpaired Treatments

p. I-10 Treatment 1: piano players.

Treatment 2: non-piano players.

a. No.

b. You are comparing results from two different treatments—paino players and non-piano players. You could consider non-players a control if desired, but a control is actually nothing more than a type of treatment.

PART 3 Defining Data Collection And Analysis Procedures (*for All Laboratories, Using or Not Using Computers*)

Example 1—Paired Treatments

p. I-11 1. Procedure: Each person measures five reaction times with his or her dominant hand (Treatment 1, observations). In the same order, each person measures another five reaction times with his or her non-dominant hand (Treatment 2, observations). These steps are repeated with the people still in the same order. There are 10 reaction times recorded per person for Treatment 1, and another 10 per person for Treatment 2.

2. Type of analysis: Because the same people are being tested under different conditions, we used a paired analysis.

a. This analysis corrects for the variability between individuals and tests the change in reaction time that occurs when each individual switches from the dominant hand to the non-dominant hand.

Example 2—Unpaired Treatments

p. I-11　1. Procedure: Each person (2 treatment groups of 2 each) measures reaction times 15 times. Students are different in the two groups so order is not critical.

　　2. Type of analysis: Because different people are being tested, we used an unpaired analysis.

　　a. This analysis randomized the individuals and allows for variability between different individuals.

EXERCISE E	The Reaction Time Experiments: Conducting the Test

p. I-12　**a.** With more replications, we can have more confidence that the average for a treatment really reflects differences between treatments rather than random fluctuations or luck. One very fast reaction time might be a chance event, but 10 fast ones in a row indicate a consistent response.

Example 1—Paired Treatments

p. I-12　Hypothesis: Reaction times for the dominant and non-dominant hand will be different.

Null Hypothesis: Reaction times for the dominant and non-dominant hand will not be different.

Prediction: The dominant hand will have a faster reaction time.

Independent variable: The hand (dominant or nondominant) used.

Dependent variable: Reaction time using each hand.

Standardized variable(s): People doing the tests, the general conditions in the lab, the number of tests for each treatment, etc.

Treatment 1: The lab group, with each person using his or her dominant hand.

Treatment 2: The same lab group, with each person using his or her nondominant hand.

Type of data analysis: Paired because the same individuals are being tested under two different conditions (dominant and non dominant hand).

Example 2—Unpaired Treatments

p. I-12b　Hypothesis: Reaction times for piano players and non-piano players will be different.

Null Hypothesis: Reaction times for piano players and non-piano players will not be diferent.

Prediction: Piano players will have faster reaction times.

Independent variable: Piano player or non-piano player.

Dependent variable: Reaction time.

Standardized variable(s): People doing the tests, the general conditions in the lab, etc.

Treatment 1: Piano players.

Treatment 2: Non-piano players

Type of data analysis: Unpaired—different individuals are being tested.

PART 1 Conducting the Reaction Time Experiments (*for Laboratories Using Computers*)
Example 1—Paired Treatments
p. I-13

Treatment	Average
(1) dominant	0.268 sec
(2) nondominant	0.273 sec

Treatment 1 Higher	Treatment 2 Higher
12	18

Chi-square = 1.2
Probability = 2.5 – 50%

Example 2—Unpaired Treatments

p. I-13

Treatment	Average	Below Median	Above Median
(1) dominant	0.493	16	14
(2) nondominant	0.472	13	17

Chi-square = 0.069
Probability = 75 – 90%

PART 2 Conducting the Reaction Time Experiments (*for Laboratories Not Using Computers*)

p. I-14   **Treatment 1 (Treatment Group 1)**

Trial	Subject							
	A	B	C	D	E	F	G	H
1	61	110	79	99				
2	45	137	61	118				
3	27	52	38	41				
4	31	70	38	44				
5	29	105	29	39				
6	63	99	27	35				
7	37	46	35	53				
8	43	143	26	34				
9	50	93	25	102				
10	88	86	24	103				
Mean	47.4	94.1	39.2	66.8				

Mean = 61.9

Treatment 2 (Treatment Group 2)

Trial	Subject							
	A	**B**	**C**	**D**	**E**	**F**	**G**	**H**
1	85	186	20	70				
2	67	145	23	74				
3	50	136	27	101				
4	96	187	33	87				
5	35	75	34	47				
6	54	107	27	97				
7	107	128	23	65				
8	121	147	41	38				
9	81	163	20	138				
10	120	120	71	91				
Mean	81.6	139.4	31.9	80.8				

Mean = 83.4

EXERCISE F   Presenting Experimental Data

p. I-16
a. Null: Men and women will have the same reaction time when country music is being played.
   Presentation format: Table because independent variable is discrete and there are only two treatments.
b. Null: The volume of rock music has no effect on the reaction time.
   Presentation format: Line graph. The independent variable is continuous and there are five treatments.
c. There are actually two experiments: a comparison of the 70 and 80 decibel reaction times with men, and then the same comparison with women. You could make it one experiment by testing the *difference* between the 70 and 80 decibel times of men against those of women, and seeing if there is a sex-based difference in the differences. Or you could treat it as a two factor experiment (sex and decibel level) and look for interaction between sex and response to decibel level.
   If treated as a decibel experiment with men followed by a decibel experiment with women, the null hypotheses will be:
   Null 1: The reaction times of men will be unaffected by an increase in noise volume from 70 to 80 decibels.
   Null 2: The reaction times of women will be unaffected by an increase in noise volume from 70 to 80 decibels.
   Presentation format: Table with 70 and 80 decibels on the top and sex down the side.
d. Again, two experiments.
   Null: The reaction times of men will be unaffected by musical style. The reaction times of women will be unaffected by musical style.
   Presentation format: Table with musical style on the top and sex down the side
e. Also, two experiments.

Null: The reaction times of men (women) will be unaffected by the volume at
which rock music is played.
Presentation format: Graph with a curve for men and a curve for women.

3. Example 1. Paired Treatment

Example 1--Paired Treatments

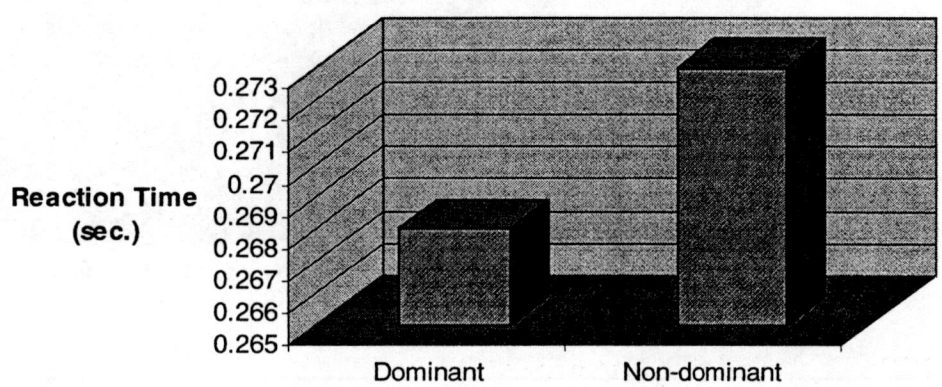

Example 2. Unpaired Treatment

Example 2--Unpaired Treatments

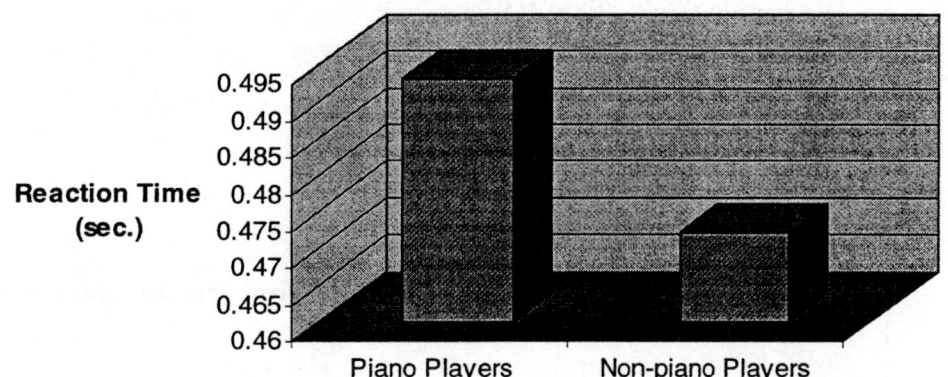

Example 3. Results for laboratories not using computers (dropping the ruler)

Example 2–Unpaired Treatments

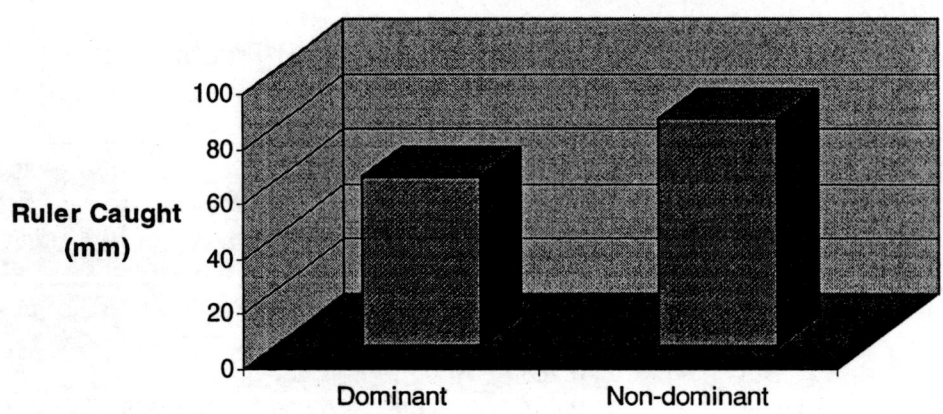

PART 1   **Analyzing Chi-Square Results (*for Laboratories Using Computers*)**

p. I-17
1. Chi-square = 1.2
   Probability = 25 – 50%
2. Degrees of freedom = 1
a. Probability is above 5%
b. We cannot reject our null hypothesis. We have no reason to doubt that the dominant and the non-dominant hand have the same reaction time.

PART 2   **Analyzing Chi-Square Results (*for Laboratories Not Using Computers*)**

p. I-18
1. Chi-square = 6.4
2. Degrees of freedom = 1
3. Probability = 1 – 2.5%
a. Probability is < 5%.
b. We must reject the null hyothesis that the dominant and the non-dominant hand
   . have the same reaction time.

PART 3   **Forming Conclusions**

p. I-19   Conclusions—in this section, students should indicate whether they accepted or rejected the null hypothesis and what this leads to: either supporting (but not proving!) their hypothesis or not supporting it. It is interesting to note that the same experiment on hand dominance provided different results and conclusions. Using different hands to punch a space bar on the computer keyboard did not show a difference in dominant and non-dominant hand use. However, when trying to catch a ruler (which requires hand-eye coordination and many more complicated finger movements), there did appear to be a difference in the use of the dominant hand vs. the non-dominant hand.

EXERCISE H   Writing a Report

p. I-19   Direct students to Appendix II.

Laboratory Review Questions and Problems

1. a. observation
   b. inference (it depends on how you define "health")
   c. inference
   d. observation
   e. It is taller (above the soil surface) and its stem is thicker but the question does not define "smaller."
   f. observation

2. World Population in 1900—6+ billion
   Title of graph: The Human Population, 1650-2000
   Dependent variable: population size.
   Independent variable: time.
   Information provided by graph: The number of people has been increasing since 1650. In recent years, this increase appears to be accelerating. The dotted line is probably a projection from about 1970...

3. a.

4. c.

5. d.

Notes:

Laboratory 1

Observations and Measurements: The Microscope

EXERCISE A Identifying the Parts of the Compound Microscope

p. 1-(2-5) **Figure lA-1 Parts of the compound microscope**
a. light source; **b.** condenser; **c.** iris diaphragm; **d.** objective lenses; **e.** turret (nose piece); **f.** ocular lens (eyepiece); **g.** body tube; **h.** stage; **i.** coarse adjustment knob; **j.** fine adjustment knob; **k.** stage clips (note: microscopes with mechanical (vernier) stages lack stage clips); **l.** base; **m.** arm.

p. 1-4 **a.** Magnification without increased resolution does not improve the perception of detail.

EXERCISE B Using the Compound Microscope

p. 1-6 **a.** Toward you.
b. Left.
c. Yes. Yes.

p. 1-7 **d.** The diameter of the field of view decreased.
e. As magnification increases, the diameter of the field of view decreases.
f. Yes. The most important habit students can learn is to continuously focus up and down to use the three-dimensional information provided by the specimen when using higher powers.
g. The total magnification is the product of the magnification of the oculars and the magnification of the objective. In this example, the magnification is 400×, using a 40× objective and 10× oculars.

EXERCISE C Preparing a Wet-Mount Slide

p. 1-9 **a.** A coverslip: 1. protects the specimen; 2. protects the objective; and 3. orients the specimen plane of focus.
b. Cyclosis transports and distributes materials such as nutrients, organelles, and metabolites, within the cytoplasm.

p. 1-10 **c.** Cyclosis slows and finally stops as cells plasmolyze.

Extending Your Investigation: Measuring Cyclosis

p. 1-10 HYPOTHESIS: Temperature affects (i.e., increases or decreases) cyclosis.
NULL HYPOTHESIS: Temperature does not affect cyclosis.
Prediction—cyclosis will increase when temperature rises.
Independent variable—temperature. Note that other factors to be considered include: light intensity; the ionic or nutrient composition of the bathing

medium; the age or health of the leaf; and the past history (pre-conditioning) of the leaf.

Dependent variable—rate of chloroplast movement (an indicator of cyclosis activity).

PROCEDURE: Measure the time (seconds) required for a chloroplast to move one micrometer (1 μm), measure the actual length of the cell (μm), and then express the rate of chloroploast movement in μm/minute). the independent variable (temperature) could be varied by immersing the *Elodea* leaf in water of the appropriate temperature (ice water, 4° C; room temperature water, 20° C; warm water, 30-35° C) prior to measuring the rate of cyclosis.

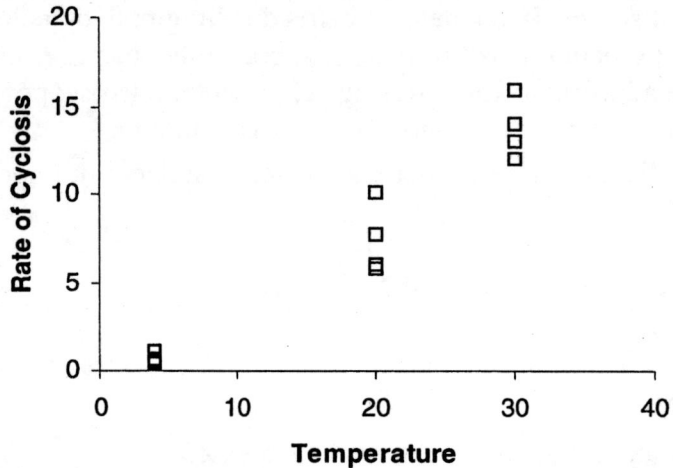

p. 1-11　　RESULTS: As temperature increased, the rate of cyclosis increased until heat killed the cells

Results support the hypothesis. Results allow the null hypothesis to be rejected. The prediction was correct.

Conclusion—temperature affects cyclosis.

EXERCISE D　　Measuring the Size of Objects Using the Compound Microscope

p. 1-11　1.　Field of view diameter is 3.8 mm = 3800 μm.

p. 1-12　2.

Objective	Diameter (μm)
4×	3800
10×	1520
40×	380
43×	353

Using a **4×** objective, the diameter of the field of view is 3800 μm (see 1 above).

Using a **10×** objective, the diameter of the field of view is 1520 μm
This value is calculated using the following equation:

(magnification A/magnification B) × diameter A (μm) = diameter B (μm)
where

---

magnification A (for the 4× objective coupled with 10× oculars) = 40×
magnification B (for the 10× objective coupled with 10× oculars) = 100×
diameter A (measured) = 3800 μm.

Therefore,

(40/100) × 3800 μm = 1520 μm.

Using a **40×** objective, the diameter of the field of view is 380 μm calculated using the equation:

(magnification A/magnification B) × diameter A (μm) = diameter B (μm)

where

magnification A (for the 4× objective coupled with 10× oculars) = 40×
magnification B (for the 40× objective coupled with 10× oculars) = 400×
diameter A (measured) = 3800 μm.

Therefore,

(40/400) × 3800 μm = 380 μm.

Using a **43×** objective, the diameter of the field of view is 353 μm calculated using the equation:

(magnification A/magnification B) × diameter A (μm) = diameter B (μm)

where

magnification A (for the 4× objective coupled with 10× oculars) = 40×
magnification B (for the 43× objective coupled with 10× oculars) = 430×
diameter A (measured) = 3800 μm.

Therefore,

(40/430) × 3800 μm = 353 μm.

3. For example, if 126 cells of *Nostoc* span a field of view having a diameter of 380 μm at 400× magnification, then each *Nostoc* cell has a diameter of (380 μm/126 cells) = 3 μm.

a. If a cell measures 10 μm at 100×, the real length of the cell at 200× is still 10 μm, though the cells will appear larger under the higher magnification.

p. 1-13 b. The value for a division on an ocular micrometer varies for each of the different objectives of the microscope because the objectives magnify the apparent size of objects.

Sample calibrations (each microscope will vary slightly):

4× objective and 10× oculars:

one division on the ocular micrometer = 2.2×10^{-2} mm = 22 μm

c. 10× objective and 10× oculars: 1 division on the ocular micrometer equals $9 \times 10^{-3} = 9$ μm.

d. 43× objective and 10× oculars: 1 division on the ocular micrometer equals 2.2×10^{-3} mm = 2.2 μm.

Sample calculations:

One *Nostoc* cell had a diameter of 2 ocular micrometer units when observed using the 43× objective and 10× oculars. Since each ocular micrometer division equals 2.2 μm, the cell diameter was calculated as:

2 ocular micrometer divisions × (2.2 μm/division) = 4.2 μm
Nostoc cell diameter

One *Paramecium* cell had a length of 8 ocular micrometer units when observed using the 4× objective and 10× oculars. Since each ocular micrometer division equals 22 μm, the cell diameter was calculated as:

8 ocular micrometer divisions × (22 μm/division) = 176 μm
Paramecium cell length

One *Chlamydomonas* cell has a diameter of 2 ocular micrometer units when observed using the 10× objective and 10× oculars. Since each ocular micrometer division equals 9 μm, the cell diameter is calculated as:

2 ocular micrometer divisions × (9 μm/division) = 18 μm
Chlamydomonas cell diameter

EXERCISE E   The Stereoscopic Dissecting Microscope

p. 1-14   Figure lE-1 The stereoscopic, or dissecting microscope
(clockwise, from top): oculars (eyepieces); body tube; objective; stage; base; arm; focus knob. (Note that a magnification changer is not indicated in this figure.)

p. 1-15   a. Reflected light works better, because 3-D specimens are usually opaque to transmitted light.

Laboratory Review Questions and Problems

1. Assume the microscope is equipped with 10× oculars: a. 40×; b. 100×; c. 430×.

2. Magnification increases the apparent size of an object. Resolution improves the perception of detail.

3. The objective is most important in determining the resolving power of a microscope. The numerical aperture (NA) is dependent upon the design of the objective. The second (uppermost) lens in an objective determines the numerical aperture, and thus, the resolving power, for this lens limits the "angular cone" of light coming from the condenser through the specimen. The larger the angular cone of light which can be utilized, the better the resolution of an object.

4. (c) Because the NA = 0.7 for this objective, its resolution would be superior. The larger the NA, the smaller the distance that can be resolved between two objects and the better the resolution.

5. Objective: determines magnification and resolution.
Ocular: increases the magnification of the objectives.
Iris diaphragm: regulates the quantity of light striking an object on a microscope stage.

6. Toward you. Toward you.

7. The condenser focuses light on an object so that the diameter of the cone of light (θ) just matches the diameter of the object to be viewed.

8. Resolution is better with a higher NA. The value R indicates the smallest distance (usually

expressed μm or nm) between two "dots" that will allow the "dots" to be distinguished as separate, rather than blending into one, i.e., to be resolved. The smaller the value of R the *greater* the resolving power of the objective.

9. $R = \lambda/2$ NA;   or 0. 204 μm.
    Note that resolution can be greatly increased by using oil with appropriate objectives.

10. 1.12 μm (1120 nm); μm (nm); 1.24 μm (1124 nm).
    Calculations: (note $n_{air} = 1$)
    Theoretical $R = \lambda/2$ NA = 0.56 μm/(2)(0.25) = 1.12 μm
    Actual R $= \lambda/2(n \sin \frac{1}{2} \theta) = 0.56$ μm/2(1 sin $\frac{1}{2} \theta$) = 0.56 μm/2(n sin $\frac{1}{2}$ (26°))
           $= 0.56$ μm/2(sin 13°) = 0.56 μm/2(0.225) = 0.56 μm/0.450 = 1.24 μm

11. The field of view is the area of the specimen that can be seen when looking through the ocular/objective combination. The size of the field of view decreases as magnification increases.

12.

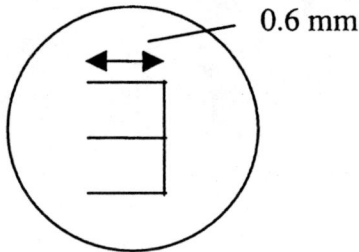

(Magnification A/Magnification B) × diameter A (μm) = diameter B (μm)
Magnification A (4× with 10× oculars) = 40×
Magnification B (10× with 10× oculars) = 100×
Diameter A = 3 mm = 3000 μm
    (40/100) × 3000 = 1200 μm
The width of the field of view is 1200 μm. The letter E takes up ½ the width of the field of view or 600 μm. The letter E is 0.6 mm wide.

13. (b) 350 μm.
    Calculation—The diagram on the left pairs a stage micrometer (on top) with an ocular micrometer (on the bottom). Since 20 ocular micrometer units equals 0.1 mm on the stage micrometer, each unit on the ocular micrometer equals 0.005 mm (= 5 μm). The diagram on the right uses the calibrated ocular micrometer to measure a cell. The cell is 70 ocular micrometer units in length. Since each ocular micrometer unit equals 5 μm, the cell length is calculated as:
        cell length (μm) = 70 ocular micrometer units × (5 μm/ocular micrometer unit) = 350 μm

Notes:

Laboratory 2

Observations and Measurements: Measuring Techniques

p. 2-2   1.

$1 \text{ nm} = 10^{-9} \text{ m}$	$25 \ \mu\text{m} = 2.5 \times 10^{-3} \text{ cm}$
$1 \text{ cm} = 10 \text{ mm}$	$10 \text{ cm} = 1 \times 10^{2} \text{ mm}$
$1 \text{ mm} = 10^{6} \text{ nm}$	$12 \text{ l} = 1.2 \times 10^{2} \text{ dl} = 1.2 \times 10^{4} \text{ ml}$
$1 \text{ m} = 10^{6} \ \mu\text{m}$	$2 \ \mu\text{l} = 0.022 \text{ ml} = 2.2 \times 10^{-2} \text{ ml}$
$1 \ \mu\text{m} = 10^{-4} \text{ cm}$	$250 \ \mu\text{m} = 2.5 \times 10^{5} \text{ nm}$
$1 \text{ mg} = 10^{3} \ \mu\text{g}$	$10 \ \mu\text{g} = 1 \times 10^{4} \text{ ng}$
$1 \text{ mg} = 10^{-6} \text{ kg}$	$9 \ \mu\text{m} = 9 \times 10^{-6} \text{ m}$
$1 \text{ l} = 10^{6} \ \mu\text{l}$	$50 \text{ g} = 5 \times 10^{4} \text{ mg}$

    a. 5 kg. Calculations: $2.2 \text{ lb./kg} \times 5 \text{ kg} = 11 \text{ lbs.}$

    b. 2 oz. Calculations: $2 \text{ oz} = 2 \times 28.53 \text{ g} = 57.06 \text{ g}$

    c. No. Calculations: $1 \text{ km} = 0.62 \text{ mile}$

                          Therefore: $85 \text{ km} \times 0.62 \text{ mile/km} = 52.7 \text{ miles}$

                          85 km/hour = 52.7 miles/hour.

p. 2-3   **d.** 2,500 ml. Calculations: $946.4 \text{ ml/qt} \times 2 \text{ qt} = 1892.8 \text{ ml.}$

    e. Yes. $50,000 \ \mu\text{l} = 50 \text{ ml.}$

    3. Conversion of numbers to scientific notation and metric units:

	Scientific Notation	Metric Unit Conversion
0.00013 g	$1.3 \times 10^{-4} \text{ g}$	$1.3 \times 10^{-1} \text{ mg}$
0.00000625 l	$6.25 \times 10^{-6} \text{ l}$	$6.25 \times 10^{-3} \text{ ml}$
2,323,000 m	$2.323 \times 10^{6} \text{ m}$	$2.323 \times 10^{12} \ \mu\text{m}$
10 μg	$1 \times 10^{1} \ \mu\text{g}$	$1 \times 10^{-8} \text{ kg}$
1654 km	$1.654 \times 10^{3} \text{ km}$	$1.654 \times 10^{6} \text{ m}$

    f. Three (3) significant digits would be used in reporting the length of line C. Both digits in front of the decimal point are significant.

p. 2-4   **g.** One. One estimated digit to the right of the decimal point, if the ruler is marked in centimeters.

    h. Five. (e.g., 100.01 g). All certain digits, plus the next estimated digit are significant digits. Using the same balance, two digits to the right of the decimal point are significant, as are the three digits to the left of the decimal point, since the accuracy of the balance is ± 0.01 g.

5. Sample Data (lengths of leaves):

Object	Length			Object	Length		
	cm	mm	m		cm	mm	m
1	7.3	73	0.073	6	5.9	59	0.059
2	5.2	52	0.052	7	7.3	73	0.073
3	7.1	71	0.071	8	7.4	74	0.074
4	7.8	78	0.078	9	7.0	70	0.070
5	7.5	75	0.075	10	7.1	71	0.071

EXERCISE B   Measuring Mass

p. 2-5 **a.** 400 g

b. 5 (e.g., 400.01 g, since the balance is accurate to ± 0.01 g, and significant digits include all certain digits (5) plus the one estimated digit specified by the ± 0.01)

3. **Table 2B-1 Measuring Mass**

Object	Mass	Possible Error	Potential Range of Values
1	4.26 g	± 0.01 g	4.25 - 4.27 g
2	8.26 g	± 0.01 g	8.25 - 8.27 g
3	4.40 g	± 0.01 g	4.39 - 4.41 g

EXERCISE C   Measuring Volume

PART 1   The Pipette

p. 2-7 **a.** 32 ml; 30 ml

b. The difference in volume is attributable to the difference in accuracy between a graduated cylinder and the pipette, combined with possible errors made during measuring and releasing 8 ml five times.

p. 2-8 **c.** 100 μl

PART 2   The Graduated Cylinder

p. 2-8 2. (a) 91 ml; (b) 90 ml; (c) 89.51 ml.

a. The volume estimated at eye level is most accurate. The volume estimated from a standing position is too high. The volume estimated from below is too low.

p. 2-9 **b.** The 10 ml cylinder would be more accurate for measuring 5 ml of solution cecause the 10 ml cylinder would spread the volume over a greater length.

Extending Your Investigation: Comparing Measuring Devices

p. 2-9 HYPOTHESIS: A beaker will be more accurate than an Erlenmeyer flask.
NULL HYPOTHESIS: There is no difference in accuracy of beakers as compared to Erlenmeyer flasks.
Prediction—measuring 50 ml will be done with more accuracy using a beaker.

p. 2-10 PROCEDURE: Measure 50 ml using a graduated cylinder. Pour into a beaker and record. Pour from the beaker into the Erlenmeyer flask and record.
Independent variable—type of measuring device (beaker or Erlenmeyer flask).

Dependent variable—amount of liquid measured.

RESULTS: beaker   slightly less than 50 ml (100 ml beaker)

can only estimate at 50 ml (50 ml beaker)

            flask   more than 50 ml

both 50 and 125 ml flasks—larger difference between bottom of meniscus and the 50 ml mark than seen with the beaker measurements)

Results support the hypothesis.

Results allow the null hypothesis to be rejected.

The prediction was correct.

Conclusion—if only a beaker and a flask are available for use in measuring liquid amounts, use the beaker because it is more accurate.

PART 3 The Volumetric Flask

a. Yes. The graduated cylinder is less accurate than is the volumetric flask. If a volumetric flask is available, always us it.

PART 4 Precision and Accuracy of Measuring Devices

p. 2-11

a. If the flask is not initially dry, the measured weight of the 100 ml of water will be inaccurate. If excess water is present on the outside of the flask, the measured weights will be greater than the true weight of 100 ml of water. If residual water is contained within the flask after taring the (presumably) empty flask, but prior to filling it with water, the measured weight will be less than the true weight of 100 ml of water, since less than 100 ml of water was added to the flask.

3. Sample data:

Flask 1: 99.31 g,

Flask 2: 99.50 g,

Flask 3: 99.50.g,

b. The range of deviations from the true value of 100g was:

$99.50 - 99.31 = 0.19$g.

(largest) − (smallest) = range

c. If we assume that 100 ml of water should weigh 100 g, then the accuracy of the volumetric flask used was ± 0.5 g, since 0.5 g was the greatest deviation from the true value.

d. One ml of water weighs 1 g only at a temperature of 4° C, where water is most dense. Since our experiments were conducted at room temperature (~20° C), 100 ml of water would be expected to weigh less than 100 g, since water is less dense at temperatures higher and lower than 4° C.

e. Other factors which could affect the accuracy of these measurements include: the presence of solutes in the water (since the given comparative values were for pure water); atmospheric pressure; and human errors.

p. 2-12

6. Sample measurements:

Measurement 1: 100.51 g

Measurement 2: 100.26 g

Measurement 3: 100.98 g

Measurement 4: 100.66 g

Measurement 5: 100.15 g

f. The range of values (largest value – smallest value) = 100.98 g – 100.15 g = 0.83 g

g. The graduated cylinder was less accurate than was the volumetric flask. The range of deviations was 0.83 g for the graduated cylinder and 0.19 g for the volumetric flask.

h. A measuring device could be inaccurate, but precise, if the range of deviations was small, but the values were not close to the true value. This could happen if the measuring device was improperly calibrated.

i. The graduated cylinder is fairly precise, allowing for repeated measurements that are close to one another although not entirely accurate. The maximum deviation from the true value was + 0.98 g.

j. Other factors that could affect the precision of measurements include: human error (in reading the measuring device or in dispensing the sample); and experimental (environmental) conditions.

EXERCISE D   Preparing Solutions

p. 2-13

a. A molar solution contains 1 mole of a substance (one gram molecular weight) in a liter of solution.

b. Obtain a mass of 111 g of $CaCl_2$ and place it into a volumetric flask. Add water to 1000 ml.

c. A molar solution contains 1 gram molecular weight of solute per liter of *solution* while a molal solution contains 1 gram molecular weight of solute per liter of solvent.

d. 11.7 g of NaCl in 1 l of solution.

e. Mass 11.7 g NaCl. Place this in a volumetric flask and add water to 500 ml.

f. Mass 29.25 g $CaCl_2$. Place this in a volumetric flask and add water to 250 ml.

1. Example—for 0.6 M NaCl, use 3.51 g/100 ml.

3. Specific gravity = 1.025

g. Assigned molarity was 0.6 and measured molarity was 0.62. As for accuracy, the solution that was prepared contained more NaCl than was needed.

h. Experimenter's error in reading the meniscus in the volumetric flask or in determining the mass for NaCl.

p. 2-14

i. For 50 ml of a 0.002 M solution, you would need 0.00585 g. Prepare the solution as a serial dilution as above, but add the final 10 ml to 40 ml of distilled water instead of 90 ml.

5. You need a solution that contains 2.34 g/100 ml.

$$58.5 \text{ g} \times 0.1 \text{ l} \times 0.4 \text{ M} = 2.34 \text{ g/100 ml}$$

A 4 M solution of NaCl contains 234 g/1000 ml, 23.4 g/100 ml, 2.34 g/10 ml. Therefore, take 10 ml of the stock solution (4 M) and add it to 90 ml of distilled water.

j. Accuracy: When compared to the standard curve, the specific gravity reading on the hydrometer was 1.015. This would indicate a solution that is 0.38 M. Since the third digit (5) is estimated, this solution is close to accurate.

1.

Leaf Size categories (cm)	Number of leaves/size category
5.0 - 5.49 cm	1
5.5 - 5.99 cm	1
6.0 - 6.49 cm	0
6.5 - 6.99 cm	0
7.0 - 7.49 cm	6
7.5 - 7.99 cm	2

Histogram of Leaf Data

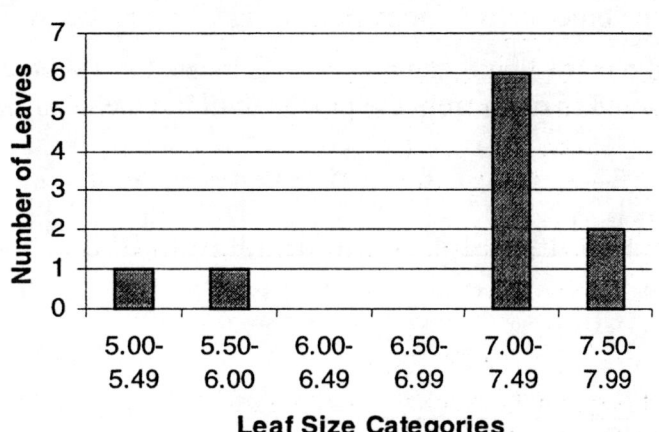

2. "Taring" is subtraction of the weight of the container from the total weight of the container plus its contents, in order to obtain the weight of the contents alone. Some (but not all) balances allow for automatic taring of the weight of the container.

3. NaOH on the balance = 60.56 g = 6.056×10^4 mg = 6.056×10^7 μg = 0.0605 kg.
   Calculations:
   > NaOH + paper = 22.21 g;
   > Since paper weighs 0.63 g, the weight of NaOH
   >> Lump #1 = 22.21 – 0.63 = 21.58 g;
   >> Lump #2 = 20.64 g;
   >> Lumps #2 & #3 = 38.98 g;
   > Since Lump #2 weighs 20.64 g,
   >> the weight of Lump # 3 = 38.98 – 20.64 = 18.34 g.
   > Total weight on the balance = 21.58 + 20.64 + 18.34 = 60.56 g.

4. Pipette #1 (TC = "To Contain" volumetric pipette) 50 μl TC at 20° C:
   > To measure out 50 μl of liquid, fill pipette to the line above the internal reservoir, then allow liquid to drain out, leaving the last drop in the pipette.

   Pipette #2 (TD = "To Deliver," blow out pipette) 25 ml in 1/10 TD:
   > Two methods may be used to measure out 12.2 ml of liquid:
   >> 1. Fill pipette to 0 and allow liquid to drain out until liquid level is at 12.2;

2. Fill pipette to 12.8 and allow all liquid to drain out, then blow out remaining drop.
Pipette #3 ((TD = "To Deliver," delivery pipette) 10 ml in 1/10 TD:
   To measure out 10 ml of liquid, fill the pipette to 0 and allow to drain until liquid level is
   at 10. Do not allow all liquid to drain out. Do not blow out.

5. 5 ml in 1/10:

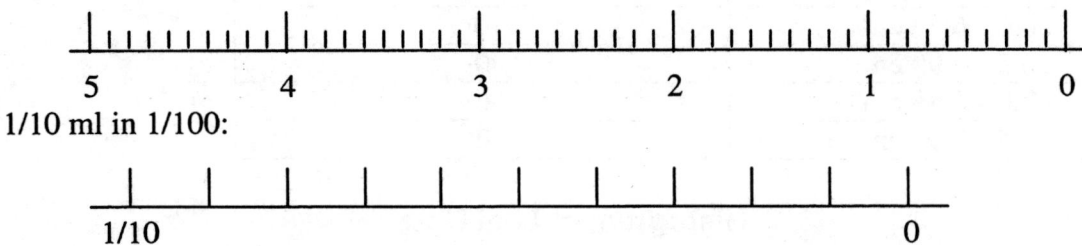

1/10 ml in 1/100:

6. No. Pipette acids and other harmful materials using a Propipette (or Pi-pump).

7. After drawing up 8.6 ml of liquid, the meniscus is located at 1.4 on the pipette. To measure
    out 6.5 ml, allow liquid to drain from the pipette until the meniscus is at 7.9.
       10 ml in 1/10

8.

	Scientific Notation	**Convert**
16 ml	1.6×10^1 ml	16,000 µl
0.0005 g	5×10^{-4} g	500 µg
150 µg	1.5×10^2 µg	150000 ng
12 nm	1.2×10^1 nm	0.000012 mm

9. The 50 ml graduated cylinder would be the normal choice. For accuracy, you could use the
    10 ml pipette, filling and emptying it two times and then dispensing the final 5 ml.

10. A volumetric flask is more accurate than a graduated cylinder for measuring volume because
    the neck of the volumetric flask extends the last few milliliters of the contents over a longer
    distance than does the comparably wider graduated cylinder. Therefore, the potential for
    inaccuracy due to reading of the meniscus level is minimized.

11. Accuracy is the degree to which a measured value corresponds to the true value. Precision is
    the degree to which measurements are reproducible when repeated.
    a. The precision of a measurement is determined by the measuring device; measuring
       technique; and the potential for experimental error.
    b. The accuracy of a measurement is determined by the experimental design; measuring
       device and technique; and human error.

12. (c) Measurements in Experiment 1 should have been closer to 5.0, but they are repeatable, so
    they are not accurate but are precise. Measurements in Experiment 2 are closer to 5 (more
    accurate) but are less precise (repeatable).

Laboratory 3

pH and Buffers

EXERCISE A   Understanding pH

p. 3-4   Table 3A-2 Calculating pH for Acids

$[H^+] M$	pH
1×10^{-3}	3
1×10^{-4}	4
1×10^{-5}	5
1×10^{-4}	4
1×10^{-2}	2
1×10^{-1}	1

p. 3-5   Table 3A-3 Calculating pH for Bases

$[OH^-] M$	$[H^+] M$	pH
1×10^{-3}	1×10^{-11}	11
1×10^{-4}	1×10^{-10}	10
1×10^{-2}	1×10^{-12}	12
1×10^{-5}	1×10^{-9}	9
1×10^{-0}	1×10^{-14}	14

3. For a neutral solution, where: $[H^+] [OH^-] = 1 \times 10^{-14}$

then: $[H^+] = 1 \times 10^{-7}$

$[OH^-] = 1 \times 10^{-7}$

pH = 7

EXERCISE B   Using Indicators to Measure pH

PART 1   Making a pH Indicator

p. 3-6

pH	Color
2	red
4	pink
6	purple
7	blue
8	blue-green
10	green
12	initially olive-green (fades to yellow)
14	initially brown (fades to gold)

PART 2   Measuring pH with Cabbage Indicator

p. 3-7   3. pH of A = 2; $[H^+] = 1 \times 10^{-2}$
4. pH of B = 6; $[H^+] = 1 \times 10^{-6}$
a. Solution A is more acidic

PART 3   Using Alkacid Test Paper

p. 3-7   2. pH of A = 2 (red)
pH of B = 6   pH of C = 2   pH of D = 6
a. Yes.
b. Cannot comment on discrepancies in basic solutions since all solutions tested were about neutral or acidic. (It is important for students to recognize that standards only work for determining values for unknowns when within the range used to develop the standards themselves.) The pH of the solutions tested (water and variously diluted HCl solutions) was accurately estimated using either the alkacid test paper or the cabbage juice indicator.

EXERCISE C   Determining the pH of Some Common Solutions (*Optional*)

PART 1   pH of Beverages

p. 3-8   HYPOTHESIS: If apple juice sweetness is caused by having less acid content, then apple juice will have the highest pH.
NULL HYPOTHESIS: The pH values will be the same for all beverages.
Prediction—apple juice will have the highest pH.
Independent variable—type of beverage.
Dependent Variable—pH.

1.

Beverage	pH
apple juice	6
coffee	4
7-Up	4
white wine	3

2. 1. White wine
2-3. Coffee and 7-Up
4. Apple juice
Results support the hypothesis. However, note that the hypothesis also suggests that the lack of acid is responsible for sweetness. This would need to be tested further to determine what is responsible for the higher pH.
Results allow the null hypothesis to be rejected.
The prediction was correct.
a. I did not think wine would be so acidic, but since I know it can go bad and taste more like vinegar, I guess that might be why.
b. People with stomach problems or any sign of ulcers should not ingest excess sodas and coffee because the low pH is very hard on the stomach lining.

PART 2   pH and Activity of Some Common Medicines

p. 3-8   HYPOTHESIS: If medicines are used to combat acid stomach, then they should

---

have high pH values.

p. 3-9   NULL HYPOTHESIS: All medicines have the same pH.

Prediction—pH of Milk of Magnesia and Maalox should be high.

Independent variable—type of medicine.

Dependent variable— pH.

Medicine	pH
aspirin	3
Milk of Magnesia	10-11
sodium bicarbonate	7.5
Maalox	10-11

Results support the hypothesis. Results allow the null hypothesis to be rejected. The prediction was correct.

a. Milk neutralizes the acid (salicylic acid) in aspirin and therefore reduces stomach irritation.

b. No. Both are acidic and could increase the acidic effect of aspirin in the stomach.

c. These medicines are basic and therefore increase the pH of the stomach.

d. If an excess of these medicines is used, you might change the pH of the stomach enough to upset the digestive process

PART 3   pH and the Action of Some Cleaning Solutions

p. 3-10   HYPOTHESIS: If soaps are made from lye and animal fat, then soaps should have high pH values

NULL HYPOTHESIS: All cleaning solutions have the same pH values.

Prediction—Tide and Cascade are harsher than Ivory and should be more alkaline.

Independent variable—type of cleaning solution.

Dependent variable—pH.

Cleaning Solution	pH
Drano	14
Ivory	6-7
Cascade	9-10
Tide	9-10

Results do not support the hypothesis but observations do not allow the null hypothesis to be accepted either. A new hypothesis needs to be formed and tested.

The prediction was not correct.

a. Most of the cleaning solutions are basic.

EXERCISE D   Soil pH and Plant Growth (*Optional*)

p. 3-11   HYPOTHESIS: If lime can be used by gardeners to neutralize acid soil, then lime must have the most alkaline pH.

NULL HYPOTHESIS: All soil and soil additives have the same pH values close to neutral.

Prediction—Lime solutions will have the highest pH.

Independent variable—soil type.
Dependent variable—pH.

Sample	pH
potting mix	6-7
clay	6
sand	7
lime	10-12
peat moss	2-4

Note: some peat moss has a pH that is close to neutral.
Results support the hypothesis; results allow the null hypothesis to be rejected.
The prediction was correct.

p. 3-12   a. Lime could be used to increase soil pH.

b. Peat moss could be used to lower soil pH.

c. Add peat moss to decrease soil pH. Low soil pH will promote uptake of aluminum ions by the plant. The aluminum will complex with the anthocyanins to prevent expression of the pink color. Therefore, the flowers will appear blue.

EXERCISE E   The pH Meter

p. 3-13   **Table 3E-1 pH of Common Solutions**

Solution	Estimated pH	Measured pH	[H+]	[OH–]
apple juice	6	6.2	6.3×10^{-7}	1.5×10^{-8}
7-Up	4	3.5	3.16×10^{-4}	3.16×10^{-11}
Maalox	10-11	10.5	3.16×10^{-11}	3.16×10^{-4}
Tide	8-10	9.6	2.51×10^{-10}	3.00×10^{-5}
Ivory Liquid	6	6.3	5.01×10^{-7}	2.00×10^{-8}

$[H^+] = 6.31 \times 10^{-4}$

a. 1.58×10^{-11}

p. 3-14   13.

Solution	pH
urine	4.23
pancreatic juice	7.10

Solution	$[H^+]$ M
lemon juice	1.58×10^{-3}
milk	3.98×10^{-7}

14.

	g/100 ml	Calculations
0.1 M NaOH	0.4	MW = 40; 1 M = 40 g/l; 0.1 M = 4 g/l
0.1 N Ca(OH)$_2$	0.37	MW = 74; 1 M=74 g/l; 1 M = 2 N; 0.1 N = 3.7 g/l
0.2 M KH$_2$PO$_4$	2.72	MW =136.1; 1 M=136.1 g/l; 0.2 M = 27.22 g/l
5% NaCl	5.0	5 g/100 ml

15.

Solution	Estimated pH	Measured pH	$[H^+]\,M$	$[OH^-]\,M$
0.1 M NaOH	10-12	12.68	2.09×10^{-13}	6.31×10^{-2}
0.1 N Ca(OH)$_2$	10-12	11.28	5.25×10^{-12}	1.9×10^{-3}
0.1 M KH$_2$PO$_4$	6	4.8	1.59×10^{-5}	4.24×10^{-10}
5% NaCl	6	6.25	5.62×10^{-7}	1.78×10^{-8}

16. $[OH^-] = 1 \times 10^{-8}\,M$
  b. 10
  c. 1×10^{-8}

EXERCISE F   Buffers

p. 3-16   **Table 3F-1 Determination of Changes in pH and Color**

	Unknown solution	Initial Color	Initial pH	Solution Added	Final Color	Final pH
A	congo red	red	8.4	0.1 N HCl	red	7.2
	thymolphthalein	colorless	8.4	0.1 N NaOH	blue	10.8
B	congo red	red	6.0	0.1 N HCl	blue	2.8
	thymolphthalein	colorless	6.0	0.1 N NaOH	blue	10.7
C	congo red	red	7.0	0.1 N HCl	red	6.6
	thymolphthalein	colorless	7.0	0.1 N NaOH	colorless	7.4
D	congo red	dark orange	4.8	0.1 N HCl	blue	3.0
	thymolphthalein	colorless	4.8	0.1 N NaOH	colorless	6.5

p. 3-17   **a.** Solution C buffered against both acids and bases. It contained both K$_2$HPO$_4$ and KH$_2$PO$_4$.
  b. Solution A buffered only against acid. It contained K$_2$HPO$_4$ only. The solution acted as a base because it "took on" H$^+$ ions, removing them from the solution.
  c. Solution D buffered only against base. It contained KH$_2$PO$_4$. The solution acts as a acid, "giving up" H$^+$ ions to complex with OH$^-$ ions from the NaOH, forming HOH (H$_2$O), so the H will not change.
  d. Solution B did not buffer against acid or base. It contained only water.
  e. Congo red. It turns blue as the pH becomes more acidic. pH 2 = blue; pH 7 = red; pH 11 = red.
  f. Thymolphthalein. It turns from colorless to blue as pH becomes more alkaline. pH 2 = colorless; pH 7 = colorless; pH 11 = blue.

p. 3-18   **Table 3F-2 Calculated Changes in H$^+$ and OH$^-$ Concentrations**

Unknown solution	Initial $[H^+]$	Final $[H^+]$	Change in $[H^+]$	Initial $[OH^-]$	Final $[OH^-]$	Change in $[OH^-]$
A 8.4	4.0×10^{-9}	6.31×10^{-8}	-5.9×10^{-8}	2.51×10^{-6}	6.3×10^{-4}	-6.27×10^{-4}
B 6.0	1×10^{-6}	1.58×10^{-3}	-1.58×10^{-3}	1×10^{-8}	5.0×10^{-4}	-4.99×10^{-4}
C 7.0	1×10^{-7}	2.51×10^{-7}	-1.51×10^{-7}	1×10^{-7}	2.51×10^{-7}	-1.51×10^{-7}
D 4.8	1.58×10^{-5}	1×10^{-3}	-9.84×10^{-4}	6.3×10^{-10}	3.16×10^{-8}	-3.097×10^{-8}

  g. Yes. Solution B does not buffer against either acid or base. It comes closest to showing a large change in H^+ or OH^- of approximately -1.58×10^{-3} [H^+] and -4.99×10^{-4} [OH^-]. All others show much smaller changes in H^+ or OH^- concentrations according to what they buffer against (i.e., solution A that buffers against acid shows a H^+ change of only -5.9×10^{-8} M when acid is added, but a much larger change of -6.27×10^{-4} when base is added. The opposite is true for solution D that buffers against addition of base. Solution C which buffers against both acid and base shows a small difference of approximately 10^{-7} for both addition of acid or base.

Laboratory Review Questions and Problems

1. [H^+] = 1×10^{-2} M; pH = 2.

2. Higher.

   Calculations: pH = 3.8 = $-$ log [H^+]
   [H^+] = 1.58 10^{-4}
   pH = 6.2 = $-$ log [H^+]
   [H+] = 6.31 10^{-7}

3. 2 pH units difference.

4. A (contains 1×10^2 more H^+).

5. The greater the H^+, the greater the acidity and the lesser the basicity.

6. 1×10^{-3} M; pH = 3.

7. NaOH + HCl $\leftrightarrow$ NaCl + H_2O.

8. 1×10^{-8} M; 1×10^{-6} M.

9.

[H^+]	[OH^-]	pH
1×10^{-8}	1×10^{-6}	8
1×10^{-4}	1×10^{-10}	4
1×10^{-3}	1×10^{-11}	3

10. Solution A [H^+] = 1×10^{-6} M; pH = 6
    Solution B [H^+] = 1×10^{-5} M; pH = 5
    F   Solution A contains 10× less H^+ ions than solution B and is, therefore, more basic
    T   Since [H^+] × [OH^-] = 1×10^{-14}
    F   Solution B contains 1×10^{-5} H^+ ions (the negative sign on the exponent is missing).
    F   Both solutions are acidic (below pH 7), although solution A with fewer H^+ ions is "more basic" than solution B
    T   Solution A is less acidic and therefore more basic than solution B
    T   You would decrease the [H^+] concentration to 1×10^{-7} [H^+] and increase the [OH^-] concentration to 1×10^{-7}
    T   For every difference in one pH unit, there is a difference of 10× in hydrogen ions; the pH scale is logarithmic

11. $[OH^-] = 1 \times 10^{-2}$ M; $[H^+] = 1 \times 10^{-12}$ M; pH = 12

12. $[OH^-] = 1 \times 10^{-1}$ M; $[H^+] = 1 \times 10^{-13}$ M; pH = 13

13.

Solution	pH
Maalox	8.51
saliva	6.71
vinegar	2.38

14.

Solution	pH	$[H^+]$ M	$[OH^-]$ M
tomato juice	4.2	6.31×10^{-5}	1.58×10^{-10}
blood plasma	7.4	3.98×10^{-8}	2.51×10^{-7}
seawater	8.2	6.31×10^{-9}	1.59×10^{-6}

15. pH is important for enzyme function and for blood to be able to load O_2 and unload CO_2.

16. A. 0.1 M Na_2HPO_4
    B. distilled water
    C. 0.1 M phosphate buffer, pH2
    D. 0.1 M NaH_2PO_4

17. Mix solutions of pH 6 and 8 together until a pH of 7 is reached.

18. The acid in the pine straw promoted the uptake of aluminum from the soil. The aluminum complexed with the anthocyanins so that their normal pink color was not expressed. If the gardener neutralizes the soil by adding lime, the hydrangeas will be pink again next year.

Notes:

Laboratory 4

Using the Spectrophotometer

p. 4-6   **a.** Molecules in solution absorb light maximally within a narrow range of wavelengths. Therefore, the absorbance of the blank is different at each wavelength, and the % transmittance must be readjusted to 100% (0 absorbance) each time the wavelength is changed.

7.

Wavelength (nm)	Absorbance	% Transmittance	Calculations $A = \log 1/T$ (or $-\log T$)
540	0.21	0.62	0.208
560	0.63	0.25	0.602
580	0.92	0.13	0.886
600	0.95	0.11	0.959
620	0.85	0.15	0.824
640	0.63	0.24	0.620

p. 4-7   8.

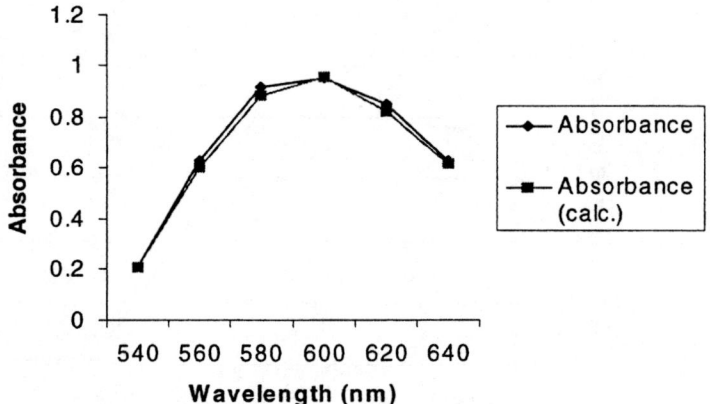

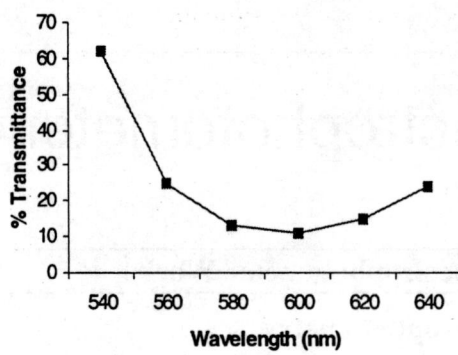

b. 580-620 nm

c. 595-600 nm

d. No. The structural characteristics of the albumin molecule determine the wavelengths of light which are maximally absorbed. These characteristics are not affected by alterations in the concentration of molecules in solution.

Extending Your Investigation: Absorbance and Transmittance

p. 4-7   HYPOTHESIS: If blue light is introduced into a red solution, then maximum absorbance of the solution will occur.

NULL HYPOTHESIS: There will be no difference in absorbance of red and blue-green or violet light by a red solution.

Prediction—absorbance by a red solution will be greatest for blue light.

Independent variable—color of light

Dependent variable—absorbance

Wavelength	Absorbance
420	1.9
470	2.0
520	1.5
570	1.1
620	0.05
670	0.01

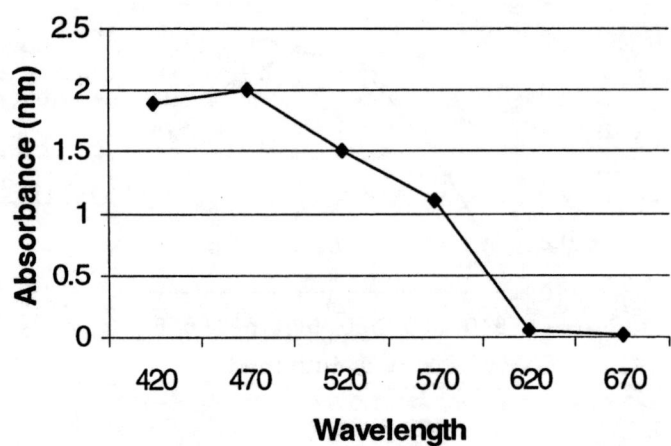

The results support the hypothesis

The results allow the null hypothesis to be rejected

The prediction was correct.

EXERCISE C   Exploring the Relationship between Absorbance and Concentration

p. 4-9 5. **a.** Each tube should contain only 0.5 ml prior to the addition of the colorimetric reagent. The last tube would contain 1 ml (0.5 ml distilled water + 0.5 ml BSA solution from the next-to-last tube), if 0.5 ml was not discarded.

6. **b.** The blank is used to readjust the spectrophotometer to 100% Transmittance (zero absorbance) and to eliminate the addition of any absorbance due to the reaction mixture contents minus the molecule (BSA) for which concentration is to be determined. The blank contains only coomassie blue + water. The blank does not contain BSA solution.

b. Serial 1:2 dilutions

8.

BSA Stock Solution	Concentration (µg/ml)	Absorbance (595 nm)
Blank	0	0
1:2	120	0.39
1:4	60	0.25
1:8	30	0.12
1:16	15	0.08
1:32	7.5	0.03
Unknown	33	0.12

9.   Absorbance of unknown = 0.12

BSA Standard Curve

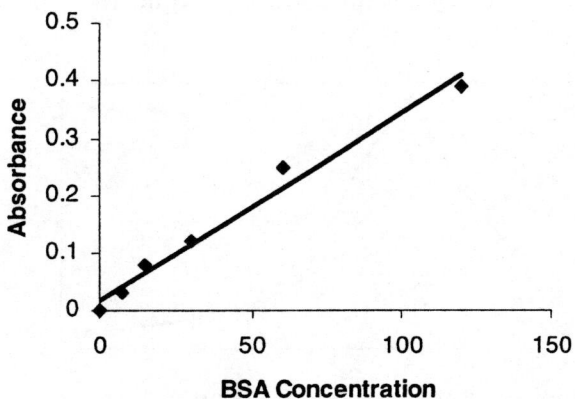

10.   Unknown: 33 µg/ml BSA

1. Green light is transmitted (and reflected) by a green solution. Red and blue light are absorbed.

2. The wavelengths of light transmitted are those "left over" after certain wavelengths are absorbed.

3. **Step 1:** Turn on the spectrophotometer and allow it to warm up for 5 minutes.
   Step 2: Select a wavelength, using the wavelength control knob.
   Step 3: With the sample holder empty and the cover on the sample holder closed, set the scale to 0% transmittance, using the zero control knob.
   Step 4: Insert the blank (clean and dry on the outside) into the sample holder. Shut the cover of the sample holder and set the Transmittance to 100% using the Transmittance control knob.
   Step 5: Insert the sample tube (clean and dry on the outside) into the sample holder. Shut the cover of the sample holder and read absorbance.
   "Zeroing" the spectrophotometer is the readjustment of the absorbance reading to 0 (100% transmittance) to assure that the absorbance readings indicate only the concentration of the molecule you wish to measure, and do not include absorbance due to unreacted molecules in the reagent itself.

4. (5, 3, 1, 4, 2, 7, 6)

5. The blank is used to readjust the spectrophotometer to 100% transmittance (0 absorbance) and to eliminate the addition of any absorbance due to the reaction mixture contents minus the molecule for which concentration is to be determined.

6. A blank would contain 5 ml H_2O, 4 ml coomassie blue.

7.
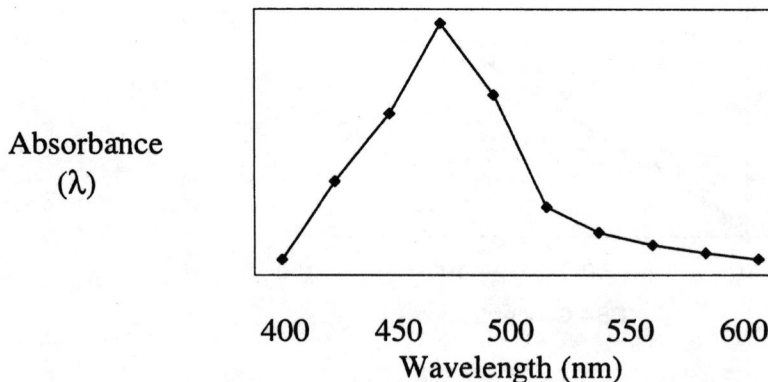

Blue light would show the maximum absorbance reading.

8. a. 510-440 nm (blue light) since it is a wavelength absorbed by a red solution.
   b. Cytoplasm extract without dye.
   c. Red dye without cytoplasm extract (use water in place of the cytoplasm extract).

d. To determine the rate of the reaction, first create a standard curve relating absorbance to known concentrations of red dye. Then mix red dye and the cytoplasm extract. Remove aliquots of this mixture and measure absorbance at defined time intervals. Determine the actual concentration of red dye present at each time by comparison to the standard curve. The rate of the reaction is equal to the change in concentration of red dye over time.

9.

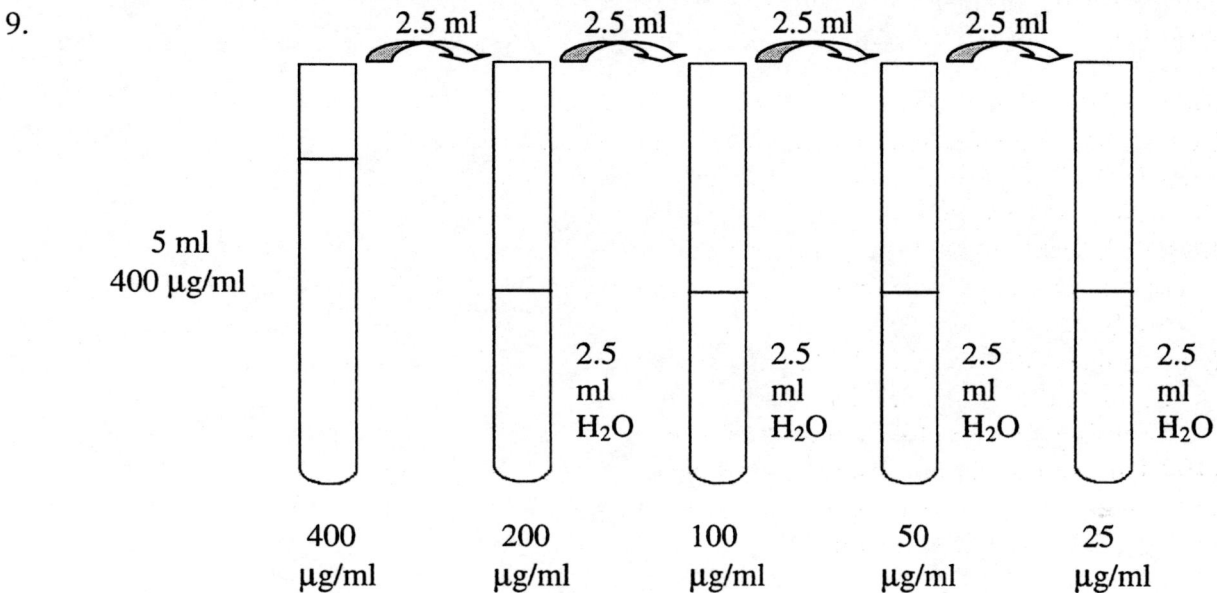

10.

Time	Absorbance	μg/ml tyrosine
0	0.4	100
0.5	0.35	80
2	0.2	50
5	0.1	25

a. 75 μg tyrosine is oxidized during the 5 minute period.
b. 25 μg tyrosine/ml/minute.
c. 8.3 μg tyrosine/ml/minute.

Notes:

Laboratory 5

Organic Molecules

EXERCISE A   Testing for Carbohydrates

PART 1   Benedict's Test for Reducing Sugars

p. 5-3   1. Water

HYPOTHESIS: If water does not contain sugar, Benedict's should remain blue.

PREDICTION: Solution will remain blue.

2. Starch

HYPOTHESIS: If starch has only one free glucose per chain, a red precipitate will not form.

PREDICTION: Solution will remain blue.

3. Glucose

HYPOTHESIS: If glucose is a reducing sugar, Benedict's solution should develop a red precipitate.

PREDICTION: Solution will contain a red precipitate.

4. Maltose

HYPOTHESIS: If maltose contains two glucose molecules that are reducing sugars, Benedict's should develop a red precipitate

PREDICTION: Solution will contain a red precipitate.

5. Sucrose

HYPOTHESIS: If sucrose is a non-reducing sugar, Benedict's should remain blue.

PREDICTION: Solution will remain blue.

6. Onion juice

HYPOTHESIS: If onions store starch like most plants, Benedict's should remain blue.

PREDICTION: Solution will remain blue.

7. Potato slice

HYPOTHESIS: If potatoes store starch like most plants, Benedict's should remain blue.

PREDICTION: Solution will remain blue.

8. Milk

HYPOTHESIS: If lactose in milk is a reducing sugar, Benedict's will develop a red precipitate.

PREDICTION: Solution will contain a red precipitate.

NULL HYPOTHESIS: All carbohydrates will react in the same manner.

Independent variable—type of substance.

Dependent variable—color reaction.

a. Water serves as the control

b. Glucose, maltose, onion juice, milk. They reacted with Benedict's reagent because they have a free aldehyde group.

c. Starch, potato, sucrose. Starch would react if it is allowed to heat long enough but, since it is a long chain of glucose molecules, there are only a few free aldehyde groups. Potatoes contain mostly starch. Sucrose has both carbon 1 of glucose and carbon 2 of fructose (the original aldehyde and ketone groups) combined in a glycosidic bond so no free electrons are available to reduce Cu^{2+}.

d. Onions store mostly sugar, so they react with Benedict's reagent. Potatoes store starch but the number of free aldehyde groups is small (see above). Milk contains lactose (glucose + galactose) which has a potentially free aldehyde group and can reduce Benedict's reagent.

p. 5-4   **Table 5A-1 Data Table For Benedict's and Lugol's Tests**

	Benedict's Test		Lugol's Test	
Tube	Original Color Before Boiling	Original Color After Boiling	Original Color Before Adding I_2KI	Original Color After Adding I_2KI
1. Water	blue	blue	yellow	yellow
2. Starch	blue	blue	yellow	blue-black
3. Glucose	blue	red precipitate	yellow	yellow
4. Maltose	blue	red precipitate	yellow	yellow
5. Sucrose	blue	blue	yellow	yellow
6. Onion juice	blue	red precipitate	yellow	yellow
7. Potato slice	blue	blue	yellow	blue-black
8. Milk	blue	red precipitate	yellow	yellow

The results support the hypotheses except for onion juice.

The results allow the null hypothesis to be rejected since solutions reacted differently with Benedict's reagent.

Predictions agreed except for onion juice.

Explanation of discrepancies—onion juice must contain reducing sugars.

PART 2   Lugol's Test for Starch

p. 5-4   1. Water

         HYPOTHESIS: If water does not contain starch, I_2KI will remain yellow.

         PREDICTION: I_2KI remains yellow.

     2. Starch

         HYPOTHESIS: If starch reacts with I_2KI, then I_2KI will turn black.

         PREDICTION: I_2KI turns black.

     3. Glucose

         HYPOTHESIS: If glucose is not a starch, I_2KI will not turn black.

         PREDICTION: I_2KI remains yellow.

     4. Maltose

         HYPOTHESIS: If maltose is not a starch, I_2KI will not turn black

         PREDICTION: I_2KI remains yellow.

     5. Sucrose

         HYPOTHESIS: If sucrose is not a starch, I_2KI will not turn black.

         PREDICTION: I_2KI remains yellow.

6. Onion juice

HYPOTHESIS: If onions store sugar instead of starch, I_2KI will not turn black.

PREDICTION: I_2KI remains yellow.

7. Potato slice

HYPOTHESIS: If potatoes store starch, I_2KI will turn.

PREDICTION: I_2KI will turn black.

8. Milk

HYPOTHESIS: If milk does not contain starch, I_2KI will not turn black.

PREDICTION: I_2KI remains yellow.

NULL HYPOTHESIS: All carbohydrates will react the same way with I_2KI.

Independent variable—type of substance.

Dependent variable—color reaction.

p. 5-5
a. Starch (and potato). Lugol's is a test specific for starch (potatoes store starch)

b. Glucose, maltose, sucrose did not react with Lugol's because they are not starches. Glucose is a simple sugar, maltose and sucrose are disaccharides, *not* polysaccharides.

c. Onion juice. No reaction; onions store sugar and not starch

Potato slice. Contains starch so it showed a positive reaction.

Milk. No reaction. Milk contains the sugar lactose but no starch

Results from each test support the hypotheses.

Results from each test allow the null hypothesis to be rejected.

Results from each test agreed with predictions.

No discrepancies were found.

d. Starch. The potato had a positive reaction to Lugol's reagent, indicating the presence of starch, and a negative reaction to the Benedict's test, indicating a lack of reducing sugars. Note that prolonged boiling of the potato slice in Benedict's reagent may produce a false positive as the stored starch is broken down into its component monosaccharides.

e. As reducing sugars. Onion juice had a positive reaction to Benedict's reagent.

EXERCISE B   Testing for Lipids

---

p. 5-6
a. The oil spot is still apparent after drying as it renders the paper somewhat translucent. The water spot vanishes (or leaves only a circumferential ring).

1. Distilled water

HYPOTHESIS: If water does not contain lipids, it should not react with Sudan IV.

PREDICTION: Water will turn pink.

2. Vegetable oil

HYPOTHESIS: If oil is non-polar, it should react with Sudan IV which is not water soluble.

PREDICTION: Red layer will float.

3. Onion juice

HYPOTHESIS: If onions do not contain lipids, onion juice should not react with Sudan IV.

PREDICTION: Solution will turn pink.

4. Hamburger juice

    HYPOTHESIS: If hamburger contains lipid, it should react with Sudan IV.

    PREDICTION: Red layer will float.

5. Cola

    HYPOTHESIS: If colas do not contain lipids, they should not react with Sudan IV.

    PREDICTION: Solution will turn pink.

NULL HYPOTHESIS: All substances react the same way with Sudan IV.

Independent variable—substance to be tested.

Dependent variable—solubility of Sudan IV

a. Oils are non-polar, whereas water is polar, therefore the oil droplets, with the incorporated Sudan IV reagent, do not mix with the water. The droplets float because oil is less dense than in water.

p. 5-7   **Table 5B-1 Data Table for the Sudan IV Solubility Test**

Substance	Sudan IV Solubility Reaction
1. Distilled Water	No reaction: Sudan IV remains in solution
2. Vegetable Oil	Positive reaction: a floating red layer (Sudan IV/oil) is formed
3. Onion juice	No reaction: Sudan IV remains in solution
4. Hamburger juice	Positive reaction: a floating red layer (Sudan IV/oil) is formed
5. Cola	No reaction: Sudan IV remains in solution

b. Lipid droplets are non-polar and do not mix with polar lipid molecules.

c. Vegetable oil and hamburger juice; these were the only substances containing fats or oils.

d. Distilled water, onion juice, cola; these substances did not contain fats or oils.

    Results support all hypotheses.

    Results allow the null hypothesis to be rejected.

    No discrepancies were found

EXERCISE C   **Testing for Proteins and Amino Acids**

PART 1   **Testing for Protein with Biuret Reagent**

p. 5-8   1. Distilled water

    HYPOTHESIS: If water does not contain protein, it should not react with Biuret reagent

    PREDICTION: Biuret reagent does not react.

2. Egg albumin

    HYPOTHESIS: If egg albumin is a protein, it should react with Biuret reagent.

    PREDICTION: Biuret reagent reacts.

3. Potato starch

    HYPOTHESIS: If potatoes do not contain protein, then potato starch should not react with Biuret reagent.

    PREDICTION: Biuret reagent does not react.

4. Glucose

    HYPOTHESIS: If carbohydrates do not react with Biuret reagent, glucose should not react.

PREDICTION: Biuret reagent does not react.

5. Amino acid

HYPOTHESIS: If Biuret reagent is a test for protein, it should not react with single amino acids.

PREDICTION: Biuret reagent does not react.

NULL HYPOTHESIS—all substances react the same way with Biuret reagent.

Independent variable—substance tested.

Dependent variable—color change.

Table 5C-1 Data Table for the Biuret Test

Substance	Color after Two Minutes	Protein Present (+) or Absent (-)
1. Distilled water	blue	+
2. Egg albumin	violet	–
3. Potato starch	blue	–
4. Glucose	blue	–
5. Amino acid	blue	–

a. Starch and glucose are not composed of proteins.

b. Water is included as a control.

p. 5-9

c. Egg albumin contains proteins (albumin)

d. All other substances did not react. Potato starch and glucose are carbohydrates and carbohydrates do not react with biuret reagent. Amino acids react with ninhydrin but not biuret reagent.

Results support all hypotheses.

Results allow the null hypothesis to be rejected.

No discrepancies were found.

PART 2 Testing for Amino Acids with Ninhydrin

p. 5-9

Table 5C-2 Data Table for the Ninhydrin Test

Solution	Final Color	Type of Molecule
A (proline)	yellow	amino acid
B (water)	no color	water
C (methionine)	purple	amino acid
D (alanine)	purple	amino acid

EXERCISE D Chromatography of Amino Acids

p. 5-12

HYPOTHESIS: Polar amino acid molecules will stay near the origin while non-polar will travel furthest with the solvent front.

NULL HYPOTHESIS: There is no difference in the tendency for different amino acids to migrate.

Prediction—lysine and aspartic acid will remain at the origin. Proline, phenylalanine, and methionine will travel furthest. Cysteine and leucine will travel intermediate distances.

Independent variable—type of amino acid.

Dependent variable—R_f (distance traveled).

p. 5-13   (sample **data**)

	Color	R_f
Sample 1:	Sample 1	
spot 1	yellow	0.75
spot 2	purple	0.67
spot 3	purple	0.60
Sample 2:	Sample 2	R_f
spot 1	purple	0.83
spot 2	yellow	0.,73
spot 3	purple	0.64

		R_f
Known amino acid		
spot 1	proline	0.75
spot 2	lysine	0.62
spot 3	cysteine	0.64
spot 4	phenylalanine	0.80
Known amino acid		
spot 1	proline	0.77
spot 2	aspartic acide	0.69
spot 3	leucine	0.94
spot 4	methionine	0.85

Indentification of Unknowns in Sample 1: proline, cysteine, lysine.

Indentification of Unknowns in Sample 2: phenylalanine, proline, lysine.

Hypothesis was correct.

Null hypothesis was rejected.

Prediction was correct

Discrepancies—none (proline turned yellow rather than purple and did not travel as far as expected.)

EXERCISE E   Analyzing Unknowns Qualitatively

p. 5-13-14   Note: the list of possible unknowns does not need to be numbered. If numbered, this list should NOT correspond to numbers on unknown tubes (this would identify the contents of the unknowns).

proline

> HYPOTHESIS: If proline is an imino (cyclic amino) acid, it will not react with ninhydrin and will turn a yellow color.
>
> PREDICTION: Ninhydrin will be yellow.

onion

> HYPOTHESIS: Since onions store sugar, the Benedict's test should be positive.
>
> PREDICTION: Benedict's will develop a red precipitate.

potato buds

> HYPOTHESIS: Since potatoes store starch, potato buds should react with I_2KI.
>
> PREDICTION: I_2KI solution should turn black.

Karo syrup

> HYPOTHESIS: Since Karo is a glucose sugar syrup, it should react with Benedict's reagent.
>
> PREDICTION: Benedict's will develop a red precipitate.

milk

> HYPOTHESIS: Since milk sugar (lactose) contains the reducing sugar glucose, it should react with Benedict's and develop a red precipitate.
>
> PREDICTION: Benedict's will develop a red precipitate.

table sugar
> HYPOTHESIS: Since table sugar is sucrose, it should not react in a positive way with any test, even Benedict's.
>
> PREDICTION: Benedict's remains blue.

egg white
> HYPOTHESIS: Since egg white is protein, it should react with Biuret reagent.
>
> PREDICTION: Biuret reagent will turn violet.

hamburger
> HYPOTHESIS: Since hamburger contains both protein and lipid (fat), it should show a positive reaction with Biuret reagent and Sudan IV.
>
> PREDICTION: Biuret reagent turns violet; hamburger juice forms a red layer with Sudan IV and water.

vegetable oil
> HYPOTHESIS: Since vegetable oil is a lipid, it should float on water as a red layer when mixed with Sudan IV.
>
> PREDICTION: Sudan IV forms a red, floating layer.

amino acid mixture
> HYPOTHESIS: If amino acids react with ninhydrin, a purple color will develop.
>
> PREDICTION: Ninhydrin will turn purple.

Table 5E-I Data Table for Analyzing Unknowns Quantitatively

Product Tested	Positive (+) or negative (−) results				
	Benedict's Test	Lugol's Test	Biuret Test	Ninhydrin Test	Sudan IV Test
1. proline	−	−	−	+	−
2. onion	+	−	−	−	−
3. potato buds	−	+	−	−	−
4. Karo syrup	+	−	−	−	− (tan)
5. milk (lactose)	+	−	−	−	− (white)
6. table sugar	−	−	−	−	−
7. egg white	−	−	+	−	−
8. hamburger	−	−	+	−	+
9. vegetable oil	−	−	−	−	+
10. amino acid mix	−	−	−	+	−

Prediction—all predictions were accurate.

Explain results—all results supported hypotheses and predictions.

Note: reactions of milk, Karo syrup, and onion are similar but observations of the colors of the unknown should assist in identification.

Laboratory Review Questions and Problems

1. Benedict's reagent complexes only with the free aldehyde or ketone groups of reducing sugars. If these groups are both involved in a bond (as between monomers of glucose and fructose, forming a α 1,2 glycosidic bond in sucrose), the Benedict's test will be negative,

even though sugars are present. All monosaccharides have a free aldehyde or ketone group which will react with the Benedict's reagent. Only some disaccharides (reducing sugars) have free aldehyde or ketone groups (those forming α 1,4 glycosidic bonds) that can reduce alkaline solutions of Cu^{2+} such as found in Benedict's reagent.

2. Biuret reagent reacts with the peptide bond of proteins. Therefore, the Biuret test is positive for proteins (dipeptides and polypeptides) but is negative for the individual amino acids which make up proteins.

3. Biuret reagent indicates the presence of whole proteins (see answer #2 above3) but not amino acids. Albumin is a whole protein.

4. Since the wax sheds water, it is probably non-polar and a lipid. When this wax is placed in a test tube with water and Sudan IV, the wax incorporates the Sudan IV to form red-colored droplets floating on the clear water.

5. Yes. Ninhydrin is yellow in the presence of proline, but the yellow color may be masked by the purple color due to reaction of ninhydrin with other amino acids in the mixture. To define the presence of proline, the component amino acids of the mixture must be separated, as by chromatography, and the isolated amino acids individually tested with ninhydrin.

6. (1) protein; (2) reducing sugar; (3) starch; (4) amino acid; (5) lipid.

7. C

8. a. III
   b. II
   c. I

9. Breakfast
Carbohydrates:	toast, cereal, pancakes, fruit
Proteins:	eggs, ham, bacon
Fats:	bacon, milk

    Lunch
Carbohydrates:	bread, fruit, pasta
Proteins:	lunch meat, cheese
Fats:	cheese

    Dinner
Carbohydrates:	rice, potatoes, pasta
Proteins:	beans, meat
Fats:	milk, butter

10. Red meat contains protein. It is also a source of saturated fats (nearly all animal fats are saturated), and cholesterol. Diets high in saturated fats are linked to health problems such as heart attacks and atherosclerosis (hardening of arteries due to formation of plaques in vessels).

11. Vitamin deficiencies are what we usually hear about: however, an excess of certain vitamins (hypervitaminosis) can also cause problems. The fat soluble vitamins A, D, E, and K should not be taken in excess because they can accumulate in the body fat and are not easily excreted. Excess vitamin A can cause loss of appetite, irritability, cracked lips, loss of hair, liver enlargement, and joint pain. Large excesses of vitamin D can cause anorexia, kidney

damage, and calcification of soft tissue. Vitamin E is less toxic, although animal studies have shown toxicity effects similar to those for A and D. An excess of vitamin K_2 appears to be associated with circulatory problems and increased breakdown of erythrocytes.

12. R_f = distance of spot from origin/distance of solvent front from origin

Known amino acids	R_f	Unknown amino acids	Distance from origin (cm)	R_f
AA1	0.25	a	2	0.25
AA2	0.66	b	5.2	0.65
AA3	0.50	c	6.4	0.80
AA4	0.80			

The mixture contains: AA1, AA2, AA4.

13. For us, essential amino acids and fatty acids are those that the human body cannot produce. The plant component of our diet serves as a source of these materials. Plants can manufacture polyunsaturated fatty acids (fatty acids containing more than one double bone).

14. The presence of unsaturated lipids in the cell membranes of plants keeps the membranes more fluid because the fatty acid chains of the lipid molecules are bent at the double bonds and cannot be packaged as tightly. Unsaturated lipids have a lower freezing point than saturated lipids. If this were not the case, the leaves and stems of herbaceous plants would freeze in the winter.

Notes:

Laboratory 6

Prokaryotic Cells

EXERCISE A   Producing Protobionts

p. 6-2
2. pH = 6.5.
5. pH = 6 (when permanently cloudy). Use 0.1 N HCl or 1% HCl (using 0.1 N HCl will shorten the procedure).
   a. The pH is probably affecting the charges on amino acids and thus causing proteins to fold and interact differently. This would also affect how proteins might fold with relation to the carbohydrates, causing them to form droplets. Hydrophobic parts of proteins would tend to interact with carbohydrates, folding them towards the interior of a sphere where water is excluded. Charged hydrophilic areas of proteins would be on the outside of the sphere where they would interact freely with water.
   b. Yes; red; red; blue.
   c. The congo red stains the coacervates red because the interior is less acidic than the exterior. The external solution turns blue. Both neutral red and methylene blue stain carbohydrates within the coacervates.

p. 6-3
   d. They expand. Although one would expect shrinkage from osmosis, there is little water inside the coacervates. Salts interact with the protein-like membranes and change their permeability properties, allowing some water to flow inward and the coacervates to expand. The added charges also cause more high molecular weight molecules to aggregate with the coacervates that have already been formed, adding to their size.
   e. Lowering the pH eventually causes disruption of coacervates because amino acids change charges and the fragile membranes break apart.
   f. Proteins. The proteins can have both hydrophobic areas (where nonpolar amino acids are located) and hydrophilic areas (where charged amino acids are present). The hydrophilic areas will interact with water while the hydrophobic areas will turn away from water. Thus, the proteins form sheets in solution. Minor pressure changes can cause these sheets to form dimples and droplets.

EXERCISE B   Examining Bacterial Cells

PART 1   Observing Bacteria Using the Light Microscope

p. 6-3
1. *Escherichia coli* are small, rod-like bacteria.
2. *Staphylococcus* are round, like small dots.
3. *Spirillum* bacteria look like small corkscrews.

p. 6-4
4. They have more than one flagellum.

PART 2 **Observing Bacteria Using the Transmission Electron Microscope (TEM)**

p. 6-5 **Figure 6B-4** *Escherichia coli* **cell in the process of dividing**
From the top: inner cell membrane, cell wall (periplasmic space), outer cell membrane, cytoplasm, nucleoid.

PART 3 **Using Gram Staining to Study Bacterial Cell Walls**

p. 6-6 HYPOTHESIS: If gram stain is used, then *E. coli* will not stain and be gram negative while *Staphylococcus* will stain and be gram positive.

p. 6-7 NULL HYPOTHESIS: Both *E. coli* and *Staphylococcus* will show the same staining properties.
  a. Prediction—*E. coli* will be pink and *Staphylococcus* will be purple.
  b. Independent variable—type of bacteria.
  c. Dependent variable—color.

p. 6-8 d. *Staphylococcus aureus* (gram positive).
  e. *Escherichia coli* (gram negative).
  f. Results support the hypothesis.
  g. No discrepancies were found.

PART 4 **Examining Cyanobacteria**

EXERCISE C **Working with Bacteria**

PART 1 **Techniques for Transferring Cultures**

p. 6-11 a. The inoculating loop is flamed before and after every transfer to destroy bacteria remaining on the loop.
  b. Placing the caps to the tubes on the laboratory bench (or any other object) would contaminate the tops, and potentially transfer extraneous bacteria to your cultures.
  c. Cool the inoculating loop before transferring cells to avoid killing (scalding) the cells.

PART 2 **Isolating Pure Cultures**

p. 6-13 a.

Streak Plate

Spread Plate

b-c. Table 6C-1 Isolating Pure Colonies of Bacteria

	Description	Organism
Streak Plate		
Colony A	white-yellow colonies	*E. coli*
Colony B	red colonies	*S. marcesens*
Spread Plate		
Colony A	white-yellow colonies	*E. coli*
Colony B	yellow colonies	*M. luteus*

p. 6-11   **d.** To prepare a pure culture of any of these bacteria, you would choose a well-isolated colony of the selected species on the plate and inoculate cells from this colony into sterile medium or agar.

Laboratory Review Questions and Problems

1. From the top: (a) cytoplasm, (b) ribosomes, (c) nucleoid, (d) inner plasma membrane, (e) outer (plasma) membrane.

2. They share many characteristics with cells including growth, limited metabolism, osmotic properties, and they can reproduce asexually (budding) but they do not contain genetic information which allows for specificity of reproduction and ability to carry information from one generation to the next.

3. Amino acids make the barrier. Hydrophobic amino acids interact with each other.

4. The bacterial chromosome is circular with only one per cell. Chromosomal proteins are absent although scaffold proteins hold loops of the chromosome together.

5. Bacterial flagella are composed of the protein flagellin. Monomer units placed end-to-end make long string-like molecules (like a string of pearls) that wrap around and around to form a hollow tube, the flagella. In contrast, the eukaryotic chromosome is composed of microtubules made from the protein tubulin (α and β tubulin dimers are made into the polymer components of microtubules). The bacterial flagellum is attached to the cell wall of the bacterium by a hook that attaches to a protein disc (in both the inner and outer plasma membranes of gram negative bacteria). The discs act like small ATP-driver "motors," whipping the flagellum around in a circle. The eukaryotic flagellum undulates from side-to-side as microtubules slide past each other.

6. The chlorophyll is found in association with the thylakoid membranes within the cytoplasm. In prokaryotes, these thylakoid membranes are NOT organized into chloroplasts.

7. You will see small, spherical pink dots.

8. Gram stain interacts with murein found in bacterial cell walls. Gram positive bacteria have very thick cell walls and thus hold the gram stain and appear purple (Gram's iodine is a purple stain). Gram negative bacteria have only a thin layer of peptidoglycan sandwiched between two plasma membranes so they do not hold the Gram stain; it washes out. To observe gram negative bacteria, a counterstain of safranin is used; it stains the cells red.

9. 1). Do not hold the culture tube up straight; always keep it at an angle so contaminants

cannot "fall into" the broth.

2). Do not put lids or caps down on the laboratory bench.

3). Always keep the lid at an angle over the Petri dish as you make a streak plate.

4). Always flame the inoculating loop and lip of the broth culture tube before making transfers.

10. To establish a pure culture from a mixed culture, inoculate a plate of nutrient agar with a small quantity of the mixed culture. Either spread or streak the inoculum onto the agar. Incubate the culture for 1-3 days, until colonies appear on the plate. Identify *E. coli* from the color and physical characteristics of the colony. Note that a precise microbiological determination of species would require plating the bacteria on agar plates containing various media to compare growth of the colonies with known growth requirements of *E. coli*. Inoculate a new agar slant or broth with cells from an isolated colony that you have identified to be *E. coli* by transferring cells using a sterile inoculating loop.

11. A. *Escherichia coli* or other non-pigmented bacteria
   B. Viral plaques
   C. *Serratia marcescens*
   D. *Micrococcus luteus*

Laboratory 7

Eukaryotic Cells

p. 7-2   3. Chloroplasts move around the cells in a uniform direction. They move across a fine network of microfilaments and microtubules (part of the cytoarchitecture of the cell). The process of cytoplasm circulation is called cyclosis and it carries the chloroplasts with it.

8. Onion cells lack chloroplasts.

a.

Similarities	Differences
cell membrane cell wall nucleus central vacuole	onion lacks chloroplasts

b. Potato cells treated with I_2KI turn blue-black. Onion cells do not stain with I_2KI (cells absorb brown color, but do not turn blue-black).

c. I_2KI stains starch grains blue-black. Thus the storage product in potatoes is starch. Onion cells did not stain blue-black because the primary storage material in onion is sugar.

d. No. Potato tubers (underground stems) typically do not develop chloroplasts unless exposed to light, and then only in the surface layer of cells. Most of the cells in the tuber are specialized for starch storage.

e. I_2KI stains starch grains blue-black. Since leucoplasts are specialized for starch storage, they turn blue-black when exposed to I_2KI.

p. 7-3   f. Banana cells contain starch when under-ripe and simple sugars when fully ripe.

g. Chromoplasts appear as faintly pigmented cells, lacking the organized thylakoid membrane system of chloroplasts.

Extending Your Investigation: Cytoplasmic Streaming

p. 7-3   HYPOTHESIS: Temperature affects the rate of chloroplast movement (cyclosis).

NULL HYPOTHESIS: Temperature does not affect cyclosis.

Independent variable—environmental conditions.

Dependent variable—rate of chloroplast movement.

PROCEDURE:

1. Place a spring of *Elodea* in ice water (4° C) for 5 minutes before observing.

2. Prepare a wet mount slide.

3. Measure the time (seconds) for a chloroplast to move 1 micrometer (μm); movement as μm/minute. (This can be done using a stage and micrometer to calibrate an ocular micrometer. If unavailable, have students use a clear plastic ruler to determine cell size and observe cells of equal size at different

temperatures. Determine the time needed for a chloroplast to travel the length of the cell.)

4. Place *Elodea* at room temperature water (20° C) and repeat observations as described above.

RESULTS:

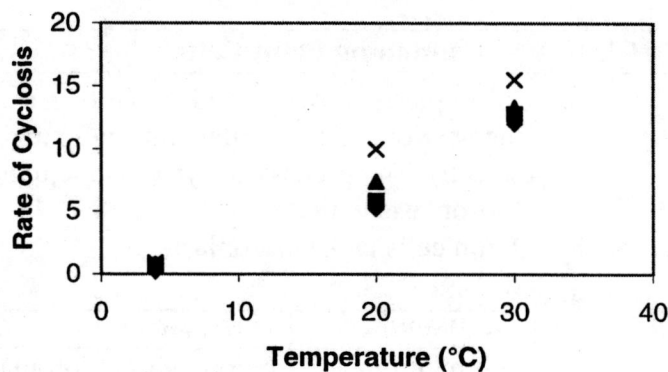

Cyclosis

Results support the hypothesis

Results allow the null hypothesis to be rejected.

Conclusion—warming the cellular environment increases the rate of cellular functions such as cyclosis.

EXERCISE B   Examining Animal Cells

PART 1   Studying Animal Cells Using Light Microscopy

p. 7-4   **a.** No cell wall. No plastids. No chloroplasts.

b.

Similarities	Differences
cell membranes	plant cells have cell walls
nuclei	plant cells have chloroplasts
mitochondria	plant cells have central vacuoles

PART 2   Studying Animal Cells Using Cytochemical Stains

p. 7-5   1. DNA: Feulgen stain:

	DNA in Mammalian Liver	Mammalian Liver Treated with DNase to Remove DNA
Color of nuclei	pinkish red	colorless to pale green
Color of cytoplasm	green	green
Other characteristics	many cell inclusions	many cell inclusions

a. Feulgen reagent stains chromosomes.

2. RNA: methyl-green-pyronin stain:

	RNA in Mammalian Liver	Mammalian Liver Treated with RNase to Remove RNA
Color of nuclei	pink	green
Color of cytoplasm	pink	green
Other characteristics	glycogen granules and blood vessels apparent	many cell inclusions (glycogen) apparent

   b. Stains RNA red (due to pyronin) and DNA plus other acidic substances green (due to methyl green).

3. Glycogen: carmine stain:

	Glycogen in Mammalian Liver	Mammalian Liver Treated with Amylase to Remove Glycogen
Color of nuclei	purple	purple
Color of cytoplasm	purple	pale purple
Other characteristics	most cells appear to have purplish cytoplasm with reddish areas of concentration on one side	granular, but cytoplasm color is uniform and does not show concentrations of red color

   c. Carmine indicates the presence of glycogen (red color) in liver cells.

4. Fat: Osmium tetroxide:

	Fat Tissue	Fat Tissue Treated with Lipase to Remove Lipid
Identifying characteristics	sac-like cells filled with black stain	black-bordered, empty cells

   d. Osmium tetroxide stains fat black.

   e. Absence of glycogen would be indicated by the purple (rather than red) color of the cytoplasm.

p. 7-6   **f.** Feulgen reagent would stain the chromosomes of *Drosophila* red.

   g. Osmium tetroxide would stain the fat globules within fat cells black.

EXERCISE C   The Strange Shape of Cells

EXERCISE D   Cell Fractionation: A Study of Eukaryotic Cells

p. 7-7   **a.** Lots of starch granules that look like small "flowers"—they appear to have a definite structure.

   b. Cell wall fragments were formed into small vesicles.

   c. Some intact cells were present, looking like small flat plates.

   d. The I_2KI stains the starch granules.

   e. Starch granules appear purple.

   f. They look like small flowers.

   g. I_2KI stains starch granules blue.

h. Pea seeds use these as a food source during germination.

i. The tube appears to have layers—a light green supernatant layer and a dark green layer above a buff-colored sediment.

j. Small, round green chloroplasts and larger nuclei are present.

k. The DPIP is being reduced by electrons that are being lost from photosystems I and II during photosynthesis (electrons eventually used to reduce $NADP^+ \rightarrow$ $NADPH + H^+$).

HYPOTHESIS: If chloroplasts are present and electrons are boosted from photosystems by photons of light during photosynthesis, then DPIP should be reduced and turn lighter in color when a chloroplast suspension containing DPIP is exposed to light.

NULL HYPOTHESIS: For solutions containing chloroplasts, photosynthesis in the presence of light will not cause the color of DPIP to differ from tubes kept in the dark.

RESULTS: **Table 7D-1**

| Tube | Color | | Absorbance | | |
	Before	After	Before	After	Difference
A	blue/green	light blue	2.0	1.5	0.50
B (dark)	blue/green	blue/green	2.0	1.9	0.01
C	blue	blue	1.5	1.3	0.02

l. The results support the hypothesis. Tube B that remained in the dark did not change in color because photosynthesis was not occurring. Tube A, however, lightened, so photosynthesis was occurring. For photosynthesis to occur, chloroplasts must be present.

m. The green layer contains chloroplasts. This test cannot verify the presence of nuclei. Also, contamination from mitochondria could occur. DPIP can also be reduced by electrons that are supposed to be transferred to NAD^+ or FAD during the reactions of the Krebs cycle and during electron transport reactions in mitochondria. In this case, the electrons have been transferred to the DPIP instead.

n. Mitochondria and smaller organelles like lysosomes.

o. Mitochondria (~1 μm) are smaller than chloroplasts (~5 μm).

HYPOTHESIS: If mitochondria are present in the supernatant then the supernatant mixed with tetrazolium should turn pink.

NULL HYPOTHESIS: The presence of mitochondria will make no difference in the color of tetrazolium.

RESULTS:

Tube	Color
X	pink/red
Y	clear
Z	clear

p. Yes. Results support the hypothesis and null hypothesis.

q. Mitochondria are smaller than chloroplasts.

p. 7-11   1. Golgi apparatus. Synthesis of secretory vesicles and their contents.
2. Centrioles (pair). Cell division.
3. Centriole. Cell division.
    a. microtubules.

p. 7-12   4. Nucleus. Cellular organization, reproduction.
    a. nucleolus; ribosome assembly
    b. nuclear pore; regulation of macromolecule exchange
    c. nuclear envelope; encloses nucleus
    d. smooth endoplasmic reticulum; lipid synthesis, carbohydrate metabolism, detoxification
5. Mitochondrion. Respiration.
    a. cristae; electron transport
    b. rough endoplasmic reticulum; manufactures membranes and secretory proteins
    c. matrix; metabolic steps of cellular respiration
6. Cell-to-cell communication
    a. plasmodesmata
    b. cell membrane
    c. middle lamella
7. Chloroplast. Photosynthesis.
    a. grana; photosynthetic electron transport
    b. thylakoid; contain photosynthetic pigments
    c. stroma; photosynthetic dark reactions

Laboratory Review Questions and Problems

1. **Animal Cell:** (clockwise, from upper right): nucleolus; mitochondrion; cell membrane; rough endoplasmic reticulum; centriole; nucleus (nuclear envelope); Golgi apparatus.
   Plant Cell: (clockwise, from upper right): chloroplast; nucleus (nuclear envelope): nucleolus; mitochondrion; dictyosome; cell wall; rough endoplasmic reticulum.

2.

Organelle(s) or Other Cell Structure(s)	In Procaryotes?	In Eukaryotes?	Function
nucleus with nuclear membrane	no (nucleoid)	yes	control center of cell; nuclear membrane compartmentalizes the cell
nucleolus	no	yes	ribosome formation
chromosomes	no	yes	genetic inheritance
cytoplasm	yes	yes	site of many metabolic pathways
mitochondria	no	yes	production of ATP
Golgi apparatus	no	yes	packaging of proteins
endoplasmic reticulum	no (mesosomes)	yes	production of protein (rough); synthesis of lipids, detoxification (smooth)
ribosomes	yes (smaller)	yes	production of protein
centrioles	no	yes	cell division
chloroplasts	no	yes (plants)	photosynthesis
cell wall	yes (peptidoglycan)	yes (plants)	structural support
flagellum	yes (simple flagella)	yes	movement

3. Feulgen stains DNA, thus it has stained the chromosomes red.
   Osmium tetroxide stains fat black. Liver does not usually (except for unusual diseased tissue) contain a lot of fat so it is unlikely that these slides should be labeled as liver.

4. Plastids are membrane-bound organelles found in plants. Chloroplasts form the photosynthetic machinery of most cells. Chromoplasts are pigmented plastids. They lack chlorophyll but synthesize and store carotenoids responsible for the red, yellow, or orange color of flowers, fruits, and leaves. Leucoplasts are colorless and contain oils, starch, and proteins. When exposed to light, leucoplasts may develop into chloroplasts. All three types of plastids were observed during the laboratory period. Chloroplasts were observed in *Elodea;* leucoplasts were observed I potato sections; chromoplasts were observed in carrot sections.

5. Yes. All of the chloroplasts moved in the same direction. Microtubules and microfilaments guide this movement.

6. The central vacuole of plant cells contains water and solutes and is surrounded by a membrane, the tonoplast. The vacuole is important for support and also increases the size of the cell as well as the amount of surface area exposed to the environment. Wastes and other storage products (such as pigments like anthocyanins) are found in the vacuole.

7. The plant cell walls, both primary and secondary, are composed mainly of cellulose and lie outside the cell membrane. Between two cells is a middle lamella composed of pectins and other polysaccharides that glue the cells together. Next to the middle lamella is the primary cell wall, composed of cellulose fibers in a matrix of glue-like polymers. Successive layers of cellulose in the primary cell wall are at right angles to one another. The secondary cell wall is found beneath the primary cell wall, next to the plasma membrane. The secondary cell wall, unlike the primary cell wall, cannot expand and is laid down after cell growth is completed.

The secondary cell wall is usually laid down in distinct layers and often contains other strengthening molecules such as lignin.

8. The cytoskeleton of a cell is composed of microtubles, actin filaments (microfilaments), and intermediate fibers. These fibers maintain the shape of the cell, anchor its organelles, direct movement of materials within the cytoplasm, and even make it possible for the cell to move. Both microtubules and actin filaments are composed of globular proteins while intermediate fibers are composed of fibrous proteins. Microtubules are involved in cell division and formation of other cellular organelles such as cilia and flagella. Actin filaments which can be assembled and disassembled rapidly are primarily involved in movement, while intermediate fibers are primarily involved in scaffolding.

9. Largest cell: starfish egg. Smallest cell: bacterium. Cells are small to maximize their surface/volume ratio.

10. Homogenization disrupts nuclear membranes and endoplasmic reticula, releasing subcelluar organelles.

11. Size and shape as well as density (determined by molecular components) determine the behavior of cellular organelles during differential centrifugation. Different speeds are used to separate major components that vary greatly in size, shape, and density and then higher speeds are used to "refine" the process. The higher speeds produce sedimentation of smaller and smaller organelles. Cell parts in our study would include:

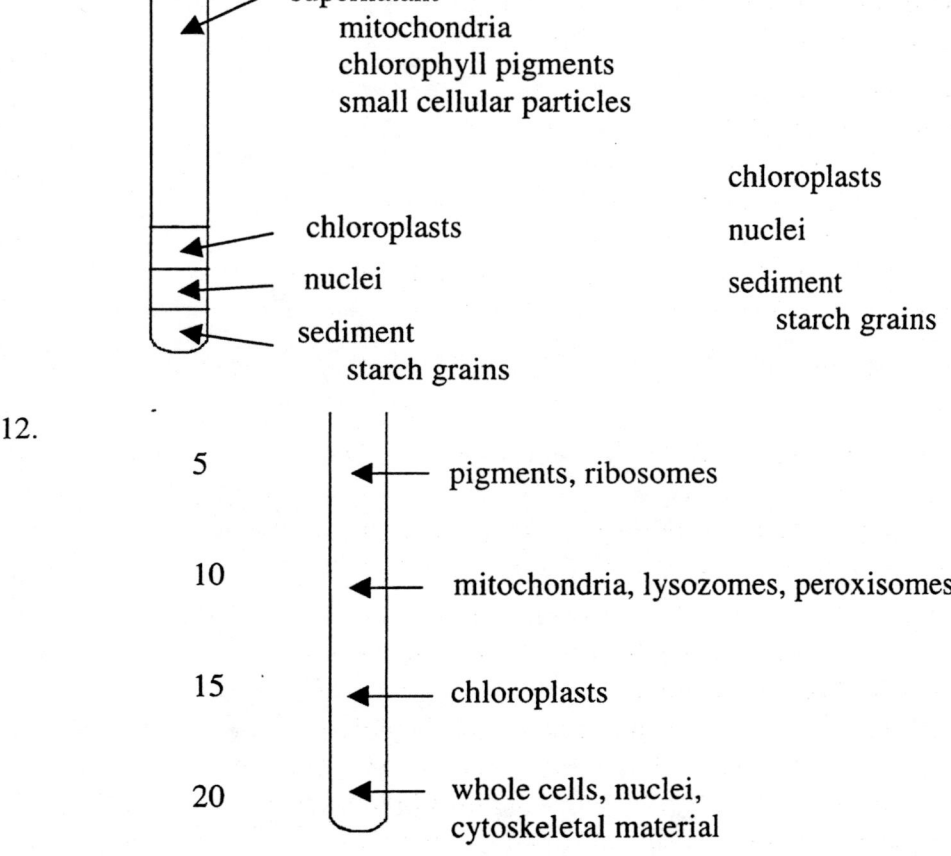

supernatant
   mitochondria
   chlorophyll pigments
   small cellular particles

chloroplasts

nuclei

sediment

chloroplasts

nuclei

sediment
   starch grains

sediment
  starch grains

12.

5   pigments, ribosomes

10   mitochondria, lysozomes, peroxisomes

15   chloroplasts

20   whole cells, nuclei, cytoskeletal material

13. DPIP and tetrazolium are both used to show reduction activity. DPIP was used to show reduction during the process of photosynthesis. Tetrazolium was used to demonstrate reduction during the Krebs cycle and electron transport in mitochondria. Both DPIP and tetrazolium accept electrons. Reduction is the addition of electrons. You can tell that DPIP has been reduced because it turns from blue to colorless with increasing reduction by electrons. Tetrazolium turns from clear to red when reduced. Reduction during photosynthesis occurs during electron transport or during the transfer of electrons to $NADP^+$ during the light-dependent reactions of photosynthesis. Reduction during cellular respiration takes place during glycolysis, the Krebs cycle, and electron transport. Since glycolysis takes place in the cytoplasm, only reduction occurring during the Krebs cycle and electron transport were measured in the experiments conducted during this laboratory.

Laboratory 8

Osmosis and Diffusion

EXERCISE A   Brownian Movement

p. 8-2   **a.** Randomly.
- **b.** No. Yes.
- **c.** Both. Water molecules and carmine molecules each have intrinsic free energy.
- **d.** Since temperature is a measure of the intensity of molecular motion, an increase in temperature increases the rate of Brownian movement.

EXERCISE B   Diffusion

PART 1   Diffusion of a Gas in a Gas

p. 8-3   **a.** The filter paper gradually turns pink.

- **b.** No.
- **c.** The chemical potential of the concentrated (liquid) solution of ammonium hydroxide in the bottom of the flask is high, thus diffusion is initially rapid and the bottom of the strip immediately becomes pink. The chemical potential of the molecules lessens as they spread out in the flask. Thus diffusion slows and the top of the paper strip only slowly becomes pink.

PART 2   Diffusion of a Liquid in a Liquid

p. 8-4   **Table 8B-1 Diffusion of a Liquid in a Liquid** (temperature 29° C)

Minute	Diameter (mm)	Radius ($D/2$) (mm)
1	10	5
2	17	8.5
3	20	10
4	20	10
5	24	12
6	26	13
7	32	16
8	36	18
9	40	20
10	44	22
11	46	23
12	50	25
13	52	26
14	52	26
15	54	27

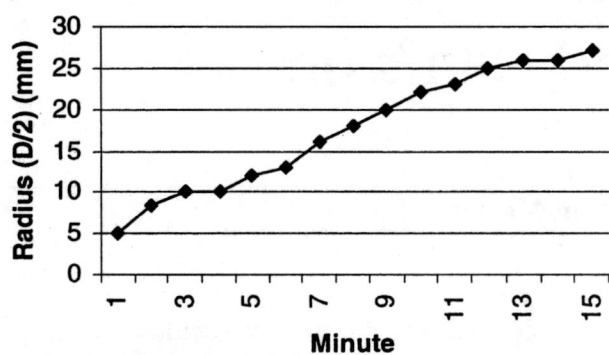

Figure 8B-1 *Graph the radius of the pigmented spot against time*

p. 8-5 **Table 8B-2 Rate of Diffusion**

Time Interval	Rate of Diffusion (mm/min)
Minute __3__ to __1__	__2.5__
Minute __8__ to __6__	__2.5__
Minute __14__ to __12__	__1.0__

a. Yes. The rate of diffusion slows as the concentration gradient decreases.
b. The chemical potential is greater for molecules closer to the center of the plate.
c. The rate of diffusion is proportional to the chemical potential gradient.
d. Yes. Net diffusion would end when the dye molecules are evenly distributed, when there is no concentration gradient.

Extending Your Investigation: Does Temperature Affect the Rate of Diffusion?

p. 8-5 HYPOTHESIS: If kinetic energy and motion of molecules increase with temperature, then diffusion rates will increase with temperature.
NULL HYPOTHESIS: Temperature does not affect diffusion rate.
Prediction—the blue dye will spread faster in hot water than in cold.
Independent variable—water temperature.

p. 8-6 Dependent variable—diameter of dye spot.

Diffusion of a Liquid in Hot Water

Minute	Diameter (*D*)	Radius (*D/2*)
1	7	3.5
2	13	6.5
3	19	9.5
4	19	9.5
5	25	12.5
6	28	14
7	31	15.5
8	35	17.5
9	38	19
10	40	20

Diffusion of a Liquid in Cold Water

Minute	Diameter (*D*)	Radius (*D/2*)
1	8	4
2	11	5.5
3	12	6
4	13	6.5
5	18	7.5
6	17	8.5
7	18	9
8	20	10
9	22	11
10	24	12

Minute	Diameter (D)	Radius (D/2)
11	40	20
12	40	20
13	40	20
14	40	20
15	40	20

Minute	Diameter (D)	Radius (D/2)
11	25	12.5
12	25	12.5
13	27	13.5
14	28	14
15	30	15

Observations—the blue dye spread faster in the hot water. This experiment was started several times. Convection currents in the hot water caused disruption of the dye until paper towels were put under the dish to avoid rapid cooling by the counter top.

Time Interval	Rate (mm/min) HOT	Rate (mm/min) COLD
Minute __3__ to __2__	3	1
Minute __8__ to __6__	1.75	0.75
Minute __14__ to __12__	0*	0.75

* Dye reached boundary of Petri dish

p. 8-7 Results support the hypothesis.
Results allow the null hypothesis to be rejected
The prediction was correct.
Conclusion—temperature affects the rate of diffusion. In the equation $G = H - TS$, anything that increases H, increases G. Also, the rate of diffusion is directly proportional to the difference in the chemical potentials between molecules at the point of origin and the area into which they diffuse.

PART 3 Effect of Molecular Weight on the Rate of Diffusion

p. 8-8 3. Distance of NH_4Cl from NH_4OH plug, $d_1 = 54$ cm.
Distance of NH_4Cl precipitate from HCl plug, $d_2 = 28$ cm.
Ratio $d_1/d_2 = 1.92$.
4. Diffusion rate for NH_3, $r_1 = 0.243$.
Diffusion rate for HCl, $r_2 = 0.165$
Ratio $r_1/r_2 = 1.47$

a. Yes (approximately, but inaccuracies may be introduced by measuring the position of the irregular plug and center of the white precipitate ring). The ratio of the distances is approximately equal to the ratio of the rates because the rate at which the molecules diffuse is related to their size; smaller molecules should diffuse faster.

b. No, because the two gases traveled at different speeds.

c. The NH_3 traveled faster. The white ring of the NH_4Cl was not seen immediately because it takes time for the two gases to move through the tube until they meet.

d. The rate of diffusion is inversely proportional to the square root of the molecular weight of the diffusing molecule.

EXERCISE C   Diffusion Across a Selectively Permeable Membrane

p. 8-9 a. The smallest molecules. Any molecule smaller than the pores in the dialysis membrane can move through.

b. Starch is a very large molecule and is probably larger than the pores in the dialysis membrane so it will probably not move through.

HYPOTHESIS: If concentrated on one side of a dialysis membrane, small molecules like glucose and I_2KI can move across the dialysis membrane, but larger molecules like starch cannot.

NULL HYPOTHESIS: Size of molecules will not affect their ability to diffuse through a semipermeable membrane.

Prediction—glucose will move out of the dialysis bag and I_2KI will move in while starch remains in the bag.

Independent variable—molecule size

Dependent variable—location of molecule at end of experiment.

p. 8-10   **Table 8C-1 Data for Dialysis Experiment**

	Original Contents	Original Color	Final Color	Color After Benedict's Test
Bag	glucose + starch	cloudy white	blue-black	orange-red
Beaker	H2O + I2KI	yellow	yellow	orange-red

c. Starch; blue color should develop

d. Reducing sugars. Benedict's reagent forms an orange-red precipitate; Tes-Tape turns green.

e. The I_2KI enters the bag, forming a blue-black color when it reacts with the starch in the bag. Glucose molecules diffused out of the bag, and could therefore be detected by the Benedict's test in the beaker. Glucose remaining in the bag reacted with the Benedict's reagent when tested.

Results support the hypothesis.

Results allow the null hypothesis to be rejected.

Conclusion—only molecules smaller than the pore size of the semipermeable membrane can pass across the membrane. Starch was too big to diffuse across the membrane. Both glucose and I_2KI were small enough to diffuse across the membrane.

p. 8-11   **Table 8C-2 Evidence: Substances Leaving and Entering**

	Outside	Inside
I2KI	yes (gold color)	yes (blue color)
Starch	no (no blue color)	yes (blue color)
Glucose	yes (red precipitate)	yes (red precipitate)

f. The red pigment is located throughout the cell and in the boiling liquid.

g. Boiling disrupts cellular membranes.

h. Boiling destroys the differential permeability of the vacuole membrane, so the pigment escapes.

EXERCISE D   A Look at Osmosis

PART 1   Measuring Osmotic Potential

p. 8-12   **a.** Negative (note: the use of the word solution indicates that solute is present in the water).

b. Yes. If enough pressure is applied to offset the negative effects of solute, then water potential can be positive.

c. Negative pressure (tension) would make water potential even lower. Water will move from an area with no solute to an area containing solute.

d. Water potential is higher outside the bag (there is no solute in the water outside the bag).

p. 8-13 e. Osmotic potential is higher inside the bag because there is more solute in the solution inside the bag when compared to an equivalent volume outside the bag.

f. Water will move into the bag because water moves from an area of higher water potential (lesser osmotic potential) to an area of lower water potential (greater osmotic potential).

PART 2 Measuring Pressure Potential: The Osmometer

p. 8-15 Table 8D-1 Data for Measuring Osmotic Potential

Contents in Dialysis Bag	Initial Mass	Final Mass	% Change in Mass
a. distilled water	17.55	17.55	0
b. 0.2 M sucrose	17.98	18.88	5
c. 0.4 M sucrose	18.11	20.18	11.4
d. 0.6 M sucrose	18.55	22.42	21
e. 0.8 M sucrose	19.37	26.31	35.80
f. 1.0 M sucrose	20.02	28.08	40.26

Figure 8D-6 *Percent change in mass of dialysis bags containing different molarities of sucrose*

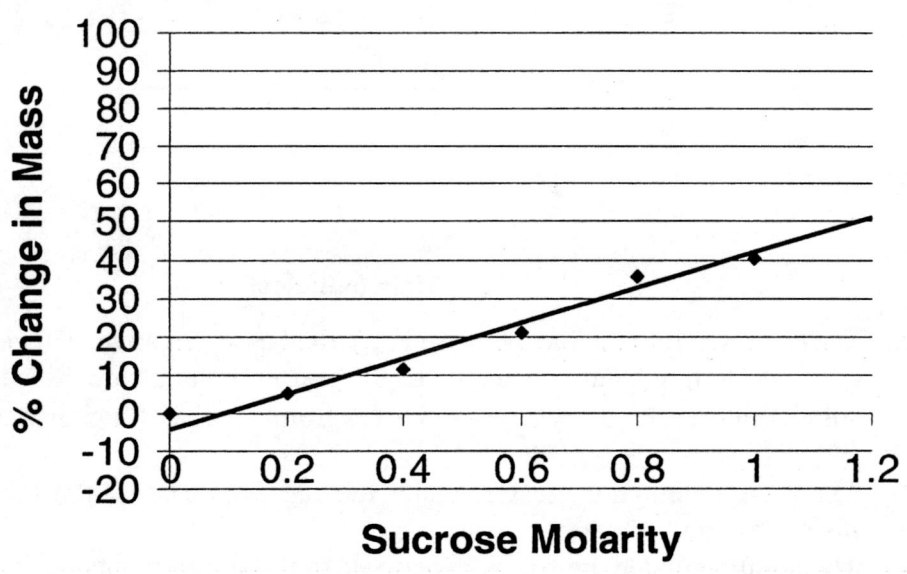

p. 8-15 a. Bags containing higher molarity sucrose solutions gained more weight.

p. 8-16 b. As sucrose concentration increases, water potential (Ψ) decreases.

c. No.

d. Hypertonic solution A has a lower water potential than does hypotonic solution B.

e. Solute concentrations of isotonic solutions are equal.

f. Bags containing sucrose concentrations greater than 0.4 M would gain weight. Bags containing sucrose concentrations less than 0.4 M would lose weight. Bags containing sucrose concentrations equal to 0.4 M would neither gain nor lose weight.

g. **Figure 8D-6** *Percent change in mass of dialysis bags containing different molarities of sucrose in 0.4 M sucrose*

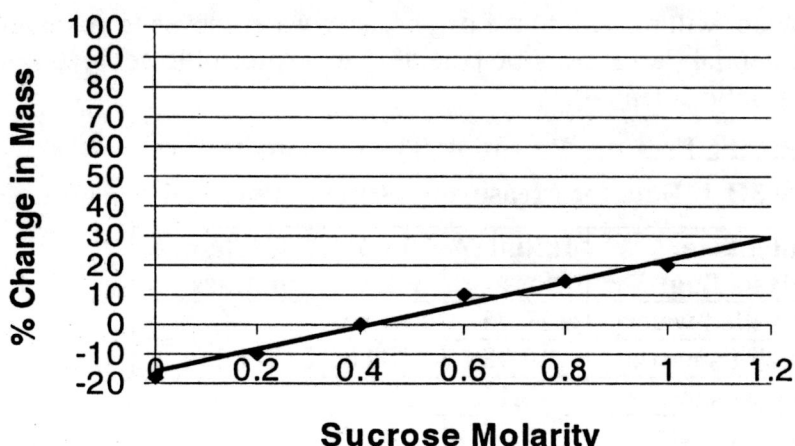

PART 2 Measuring Pressure Potential: The Osmometer

p. 8-17 **Figure 8D-8** *Graph osmotic pressure in millimeters of solution traveled up the osmometer tube*

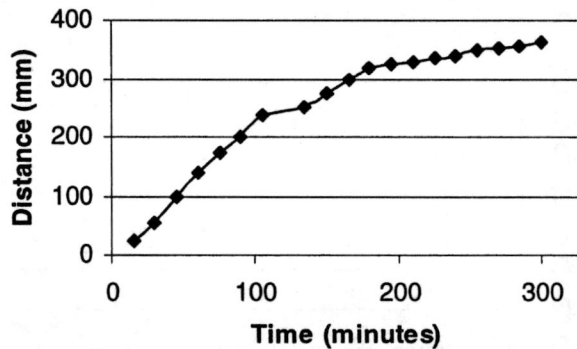

a. Water moves into the bag because Ψ_{inside} the bag is negative. This is because the concentration of solute inside the bag is greater than outside. Since there is no solute outside, $\Psi_{outside} = 0$. Water moves from an area of higher water potential (Ψ) to lower water potential.

b. The water potential is greater outside the bag. Solute concentration is higher inside the bag.

c. The solution inside the bag is hypertonic to the solution outside. The solution outside the bag is hypotonic to the solution inside the bag.

d. Osmotic pressure is exerted by the movement of water into the bag due to the greater water potential outside the bag. In an osmometer at equilibrium, the pressure exerted on the walls of the osmometer by the column of fluid that has risen in the tube is hydrostatic pressure caused by the weight of the solution. (This actual pressure, which is proportional to solute concentration, is the same as would be exerted on the wall of the osmometer if the solute particles existed as a gas of the same volume). By applying this quantity of pressure, or osmotic pressure, the tendency of the water to enter the osmometer by the process of osmosis is negated.

e. The build up of pressure inside the bag raises the water potential of the solution inside the bag.

f. When the fluid stops rising in the glass tube, the water potential inside the bag is equal to the water potential outside the bag. In this case, since $\Psi_{outside} = 0$ and the sucrose molecules in the bag are too large to move through the bag, then enough pressure must build up inside the bag for the Ψ_{inside} to also be $\Psi_{inside} = 0$.

p. 8-18   g. Net movement of water will stop when $\Psi_{inside} = \Psi_{outside}$, but the movement of individual water molecules in and out of the bag continues; a dynamic equilibrium has been established between the solutions inside and outside the osmometer bag.

PART 3   Measuring the Water Potential of Living Plant Cells

p. 8-19   HYPOTHESIS: Turnips will have a higher water potential than white potatoes.
NULL HYPOTHESIS: All types of starchy vegetables have the same water potential regardless of carbohydrate content.
Prediction—Turnips will gain more mass than white potatoes.
Independent variable—molarity of sucrose solutions that plant tissue is soaked in.
Dependent variable—% gain in mass by plant tissue.

Table 8D-2 Group Data (white potato)

Contents of Beaker	Initial Mass	Final Mass	% Change in Mass
a. distilled water	4.07	4.96	+ 21.87
b. 0.2 M sucrose	3.59	3.76	+ 4.74
c. 0.4 M sucrose	3.76	3.37	– 10.37
d. 0.6 M sucrose	3.97	3.05	– 23.17
e. 0.8 M sucrose	4.26	2.91	– 31.69
f. 1.0 M sucrose	4.25	2.81	– 33.88

p. 8-20   **Figure 8D-11** *Percent change in weight of potato cores at different molarities of*

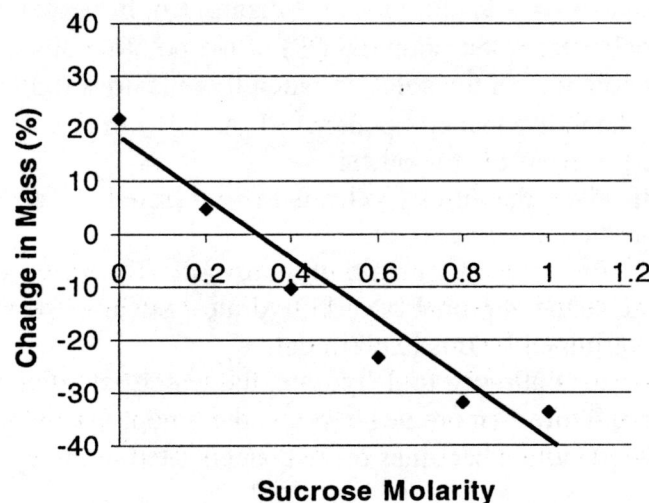

sucrose

Table 8D-2 Group Data (turnips)

Contents of Beaker	Initial Mass	Final Mass	% Change in Mass
a. distilled water	2.48	3.30	+ 33
b. 0.2 M sucrose	2.41	2.90	+ 20
c. 0.4 M sucrose	2.56	3.36	+ 18
d. 0.6 M sucrose	2.91	3.10	+ 6.5
e. 0.8 M sucrose	2.98	3.00	+ 0.67
f. 1.0 M sucrose	3.05	3.02	− 0.98

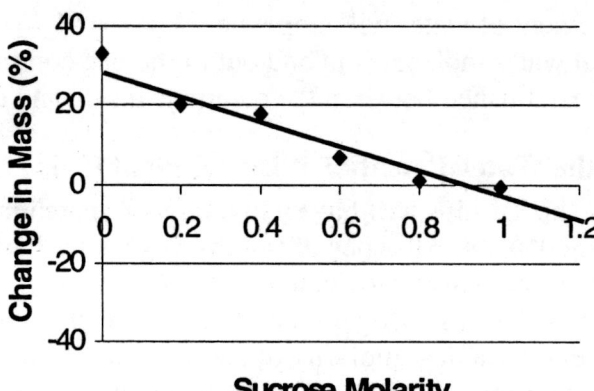

13. 0.08 M

p. 8-21

a. At equilibrium, there is no net flow of water because solute concentrations in the buffer and in the potato cells are equal.

$\Psi_{\text{white potato cells}} = -6.46$ bars

$\Psi_{\text{turnip cells}} = -22.36$ bars

$\Psi = -iCRT = -(1) \times (0.26) \times (0.0831) \times 299° = -6.46$ bars

$\Psi = -iCRT = -(1) \times (0.9) \times (0.0831) \times 298° = -22.36$

b. No. Since the sucrose solution is an open system, $\Psi_p = 0$. Therefore, $\Psi = \Psi_\pi$. But in the potato cells, there is some pressure. Thus, $\Psi \neq \Psi_\pi$. For instance, a Ψ of − 6 could be from a Ψ_p of + 2 and a Ψ_π of − 8 ($\Psi = \Psi_p + \Psi_\pi$; $\Psi = +2 - 8$) or from $\Psi_p = +3$ and $\Psi_\pi = -9$, etc. Since we cannot (in this experiment) measure Ψ_π, we can only calculate water potential (Ψ) of the potatoes since it must be equal to the water potential of the solution when there is no weight increase; we simply do not know the individual components (Ψ_p and Ψ_π) of Ψ.

Results support the hypothesis.

Results allow the null hypothesis to be rejected.

The prediction was correct.

Conclusion—turnips contain more osmotically-active solute than white potatoes. The solute is probably carbohydrate—sugar or starch—but this cannot be determined by this experiment.

c. If a potato is allowed to dehydrate, the water potential of its cells would become lower and lower (more negative) as the amount of solute/volume of solution increased (solute becomes more concentrated in the cytoplasm).

Cumulative class data for the week can be used to perform a chi-square median test. (See *Instructor's Manual: Preparator's Guide* for Laboratory I, p. I-8.)

	Average	Maximum	Minimum
red potato	– 16.87	– 9.96	– 29.88
beet	– 19.29	– 11.95	– 35.86

Chi-square median value = 2
Probability = 10 – 25%

Based on the chi-square value of 2 with *df* = 1, the probability of 10=25%, we cannot reject the null hypothesis. There is not enough evidence from these experiments to indicate that red potatoes and beets show a significant difference in water potential.

Extending Your Investigation: Water Potential of Different Tuber Types

p. 8-22   The probable cause for the data not supporting the student's hypothesis is that carbohydrate stored by sweet potatoes, primarily starch, is found in leucoplasts and is less accessible and osmotically active. Sugar stored in beets (sugar beets) and turnips is more osmotically active. Of course, the data may be in error. Have your students check out this data if they do not believe it or cannot offer an explanation.

PART 4   Observing Osmosis in a Living System

p. 8-23   a. The plant materials (carrots, celery, *Elodea* leaves) in pure water are turgid. The plant materials in salt water are flaccid and the individual cells are plasmolyzed.
            b. Yes. Yes
            c. No. The pressure exerted by the plant cell wall inhibits imbibition of excess water, therefore plant cells will stop taking in water before bursting.
            2. Chloroplasts are located at the periphery of the cell because the center of the cell is occupied by a large, fluid-filled vacuole.

p. 8-24   d. Water left the cell, consequently the central vacuole diminished in size. The chloroplasts will appear to be in the center of the cell, since the vacuole is not present to push the cytoplasm, including the chloroplasts, towards the periphery.
            e. If the cells remain viable, they will rehydrate when the salt solution is replaced by water.

Laboratory Review Questions and Problems

1. **a.** a   **b.** b   **c.** c   **d.** c   **e.** c   **f.** a

2. Salt and sugar prevent growth of bacteria and fungi by causing osmotic withdrawal of water from any contaminating organisms.

3. The potatoes are placed in a hypotonic solution (water) to cause them to become maximally turgid, and therefore to keep them firm (turgid) during the frying process.

4. In order for animal cells to neither burst or crenulate from osmotic shock, fluids given intravenously should be isotonic to body fluids. If an injection is hypertonic, cells would crenulate. If an injection is hypotonic, cells would burst. If you change the ionic content of

the blood, you change the osmotic properties of the membranes of the cells that compose the vessel walls. The kidneys will try to excrete excess ions if they are salts, but this is effective only up to a certain point.

5. The high concentration of solutes (fertilizer) in the soil caused the roots to lose water to the soil.

6. Water will enter the plant cell and pressure will build up until the water potential inside the cell is equal to the water potential of pure water outside the cell (zero).

7. Water will enter the animal cell and it will swell until it bursts. There is no rigid cell wall to help in building up osmotic pressure to offset the difference in water potentials between cytoplasm inside the cell and water outside the cell.

8. Bag $= -35$ bars $= \Psi_\pi + \Psi_p$

    where: $\Psi_p = 0$

    therefore: $\Psi_{bag} = -35 = \Psi_p + 0$

    therefore: $\Psi_{bag} = -35 = \Psi_\pi$

    If the osmometer is placed in a beaker of water ($\Psi = 0$), then water will flow into the bag ($\Psi_{bag} = -35$) until $\Psi_{bag} = \Psi_{outside}$.

    Assume: $\Psi_{\pi(outside)}$ always $= 0$ (since sucrose cannot leak out of the bag)

    $\Psi_{p(outside)}$ always $= 0$

    $\Psi_{outside} = 0$ (since $\Psi = \Psi_\pi + \Psi_p = 0 + 0$)

    Therefore, at equilibrium:

	Ψ	Ψ_p	Ψ_π
Bag	0	-17.5	-17.5
Beaker	0	0	0

    Note: equilibrium occurs when there is no net movement of water across the membrane (when $\Psi_{bag} = \Psi_{outside}$)

9. $Y_{cell} = \Psi_{\pi(cell)} + \Psi_{p(cell)}$ $\Psi_{outside} = 0 + 0$ (pure water)

    where: $\Psi_{\pi(cell)} = -4$

    $\Psi_{p(cell)} = +2$

    therefore: $\Psi_{cell} = -4 + 2 = -2$

    a. Water will diffuse into the plant cell (since $\Psi_{cell} = -2$; $\Psi_{outside} = 0$)
    b. Net diffusion will stop when $\Psi_{cell} = \Psi_{outside}$ (i.e., when $\Psi_{cell} = 0$)
    c. At equilibrium, $\Psi_{cell} = 0 = \Psi_{\pi(cell)} + \Psi_{p(cell)}$.
    If we assume that $\Psi_{\pi(cell)}$ does not change, then $\Psi_{\pi(cell)} = -4$.
    Then: $\Psi_{cell} = -4 + 4$

    $\Psi_{p(cell)} = +4$ (the cell has become more turgid)

10. outside solution: 0.5M sucrose; 27° C

    protozoan cell: $\Psi_{\pi(cell)} = -2$

    $\Psi_{p(cell)} = 0$

    $\Psi_{p(outside)} = -iCRT = -(1)\times(0.5)\times(0.0831)\times(300) = -12.47$ bars

    where: $i = 1$ (ionization constant for sucrose)

$$C = 0.5M$$
$$R = 0.0831$$
$$T = 273 + 27 \text{ (at } 300°K)$$

$$\Psi_{cell} = \Psi_{\pi(cell)} + \Psi_{p(cell)}$$
$$\Psi_{cell} = -2 + 0$$
$$\Psi_{cell} = -2$$

a. Water will diffuse out of the cell ($\Psi_{outside} = -12.47$; $\Psi_{cell} = -2$).

b. Net diffusion will stop when $\Psi_{outside} = \Psi_{cell}$.
   Since the cell volume is inconsequential compared to the volume of the outside solution (therefore the quantity of cell water lost to the outside will not appreciably change the $\Psi_{outside}$), net diffusion will stop when $\Psi_{cell} = -12.47$.

c. Cell will be plasmolyzed, since it has lost water.

d. At equilibrium, Ψcell = Ψoutside = -12.47.
   Since: $\Psi_{p(cell)}$ always = 0 (protozoan cells lack cell walls)
   Then: $\Psi_{cell} = -12.47 = \Psi_\pi + 0$
   Therefore: $\Psi_\pi = -12.47$.

11. outside: $0.2M$ NaC1; 27° C
    plant cell: $\Psi\pi(initial) = -8$, $\Psi_{p(initial)} = +2$
    $$\Psi_{cell} = \Psi_{\pi(cell)} + \Psi_p = -8 + 2 = -6$$
    $$\Psi_{outside} = -iCRT$$
    where: $i = 2$ (since NaC1 dissociates into two ions, Na^+ and Cl^-, in water)
    $$C = 0.2M$$
    $$R = 0.0831$$
    $$T = 273 + 27 = 300° K$$
    $$\Psi_{outside} = (-2) \times (0.2) \times (0.0831) \times (300) = -9.97$$

    a. Water will diffuse out of the cell (since $\Psi_{outside} = -9.97$)

    b. At equilibrium, $\Psi_{cell} = \Psi_{outside} = -9.97$

    c. As the cell becomes less turgid, $\Psi_p \rightarrow 0$. Therefore, $\Psi = \Psi_\pi + 0 = -8 + 0 = -8$.
       The cell will lose water until $\Psi_\pi = -9.97$.

    d. The cell is less turgid at equilibrium.

12. Animal cell: 1 ml volume; $\Psi_\pi = -5$ bars; $\Psi_p = 0$
    outside: $0.8M$ sucrose; 27° C
    $$\Psi_{outsice} = -iCRT = -(1) \times (0.8) \times (0.0831) \times (300) = -19.94$$
    where: $i = 1$
    $$C = 0.8$$
    $$R = 0.0831$$
    $$T = 273 + 27 = 300° K$$
    $$\Psi = \Psi_\pi + \Psi_p = -5 + 0 = -5$$

    a. Water will diffuse out of the cell ($\Psi_{outside} = -19.94$; $\Psi_{cell} = -5$).

    b. At equilibrium, Ψcell = Ψoutside = -19.94
       Assume: $\Psi_{outside}$ does not change appreciably due to the comparatively much greater volume of water in the beaker, in comparison to the much smaller (1 ml) volume of the plant cell.

    c. Cell volume = ($\Psi_\pi \times$ Volume$_{initial}$) / $\Psi_{\pi(final)} = ((-5) \times 1) / -19.94 = 0.25$ ml

Notes:

Laboratory 9

Mitosis

EXERCISE A   The Cell Cycle—Interphase and "Getting Ready"

p. 9-2   **a.** Doubling of cell size, enzymes and organelles provides duplicate materials to supply the two daughter cells which are formed by cell division.

p. 9-3   **b.** DNA must duplicate in order to provide each new cell with a complete copy of the genetic information.

p. 9-3-4   **c.** G_1 phase: cell doubles in size; organelles double in number; enzyme synthesis; centrioles begin duplication.
S phase: DNA replication.
G_2 phase: spindle fibers assemble; centrioles continue to divide and complete duplication.

EXERCISE B   Simulating the Events of Interphase, Mitosis, and Cytokinesis

p. 9-5   **a.** Each chromatid is composed of a double-stranded DNA molecule (2 polynucleotide strands, hydrogen-bonded to form a Watson-Crick double helix).

b. Chromosomes exist as homologous pairs. One comes from the father (paternal) and the other from the mother (maternal) at the time of fertilization. The two colors represent maternal and paternal chromosomes carrying the alleles specified for that chromosome.

Two red strands would signify that the two homologs had the same origin (both maternal or both paternal). Unless nondisjunction occurs the chromosomes must be of different colors. Likewise, different lengths would indicate two *different* chromosomes carrying different sets of genes and these would not be considered to be homologous.

c. Centrioles occur in pairs. One pair must move to each side of the cell to form the spindle apparatus for mitosis.

p. 9-7   **d.** Chromosomes line up in single file so that ALL chromosomes have their chromatids separated to give rise to two identical cells—the product of mitosis.

p. 9-8   **e.** Chromatids become chromosomes when each has its own centromere region, unshared by the other chromatid.

f. One DNA helix (double-stranded Watson-Crick helix) is present in each daughter chromosome.

g. We started with two (one yellow and one red) and now we have two cells, each with two (one yellow and one red) chromosomes. Thus, cell *duplication* has taken place.

p. 9-9   **h.** The number of chromosomes in each daughter cell is the same as the number present in the parent cell; mitosis and subsequent cell division (cytokinesis) duplicate the parent cell.

i. Each daughter cell contains chromosomes identical to those in the parent cell.

p. 9-10   **j.** No

k. One chromosome of each homologous pair in the daughter cell is maternal and one is paternal.

l. Mitosis maintains a constant chromosome number by: 1) duplicating each chromosome prior to cell division; and 2) segregating half of each duplicated chromosome into one of the two daughter cells. Daughter nuclei are genetically identical because they contain halves of duplicated chromosomes, thus the same amount and kind of genetic material as was present in the parent cell

EXERCISE C   Mitosis in Living Tissues—Onion Root Tips

p. 9-11   Functionally mature cells occupy the region of the root above the region of differentiation.

a. Protection of the root meristem as the root pushes through the soil.

EXERCISE D   Phases of the Cell Cycle in the Onion Root Tip

p. 9-12   **Table 9D-1 Percent of Cells in Each Phase of the Cell Cycle**

	Field 1	Field 2	Field 3	Total	% Grand Total (total/ grand total × 100)
Interphase	127	143	151	421	89.3%
Prophase	21	2	2	2	5.3%
Metaphase	2	5	4	11	2.3%
Anaphase	3	3	1	7	1.5%
Telophase	1	3	3	7	1.5%
Grand Total				472	

3. Time spent in: prophase   0 hr   51 min

metaphase   0 hr   23 min

anaphase   0 hr   14 min

telophase   0 hr   14 min

Total time spent in mitosis   1 hr   42 min

Time spent in interphase   14 hr   18 min

Note: students must be viewing the apical meristem region. If they move up toward the region of elongation, the percentage of interphase cells will be much higher and will result in a miscalculation for the length of the interphase.

a. Mitosis: 10.7%. Interphase: 89.3%

b. The results generally agree. Prophase was the longest phase; anaphase and telophase the shortest; and metaphase intermediate in duration. Most of the 16 hour cycle was spent in interphase.

c. Dependent upon individual results. If an overabundance of cells was found in interphase, the cell cycle would have to be longer if times for M phases are to agree with known data.

p. 9-13   **d.** A longitudinal section exposes the growing (elongating) axes of the cells. Thus root meristem cells will be cuboidal whereas cells in the elongation and differentiation regions will be columnar.

    e. Some cells appear empty because the section cut through the central vacuole.

Laboratory Review Questions and Problems

1. From the left: G_1; S; G_2; Mitosis; G_1.

2.

	Number of chromatids per chromosome
G_1	1
S	2 (at end of S)
G_2	2
Prophase	2
Metaphase	2
Anaphase	1
Telophase	1

3. The cell cycle is regulated by a set of proteins that interact at certain "checkpoints" in the cell cycle. There are two major types of proteins that interact—cyclin-dependent protein kinases (CdKs) and cyclins.

    Cyclin-dependent protein kinases are enzymes that phosphorylate other enzymes or proteins such as those associated with microtubules that form the mitotic spindle. Phosphorylation by kinases causes the molecules involved to complete reactions that are necessary to carry the cell cycle beyond a checkpoint. Kinases (CdKs), however, do not act alone. They must bind to cyclin proteins. Different cyclins regulate the actions of individual kinases and, with each "turn" of the cycle, the cyclin proteins are destroyed and resynthesized.

4. In multicellular organisms, cell division increases the size of the organism and replaces dead or sloughed off cells. In unicellular organisms, cell division is a method of asexual reproduction.

5. The centromere region attaches homologous chromosomes together during division.

6. The centriole is believed to function in organizing the spindle. The region surrounding the centrioles is the microtubule organizing center (MTOC).

7. Spindle fibers are involved in separating sister chromatids during mitosis.

8. A chromatid becomes a chromosome when it has its own centromere.

9. Eight chromosomes (4 yellow and 4 red), each composed of two chromatids, would be lined up in single file at metaphase of mitosis. Each daughter cell would have 8 chromosomes.

10. Make one long red chromosome and one long yellow chromosome (same number of beads on the two sides of the centromere as you used for the red chromosome). Make one short red and one short yellow chromosome that look identical except for color. Note that all chromosomes should be single strands since no mention has been made that the cell is dividing.

11. Yes. A haploid cell can undergo mitosis to form two daughter cells, each haploid cell with 4 chromosomes. At metaphase all 4 chromosomes would be lined up single file in the center of the cell.

12. Prophase
    Interphase (G_1-G_2)
    Metaphase
    Anaphase
    Interphase (G_2)
    Telophase
    Interphase (S)
    Prophase
    Anaphase.

Laboratory 10

Enzymes

EXERCISE A   Investigating the Enzymatic Activity of Catecholase

PART 1   The Effect of Temperature on Enzyme Activity

p. 10-2   **a.** The secondary structure represents the folding of the enzyme's polypeptide chain in two planes. Usually the structure such as an α helix or β sheet is held together by hydrogen bonds between amino acids. The tertiary structure represents the folding of an enzyme's polypeptide chain in three dimensions. Usually hydrogen bonds and disulfide bonds are responsible for this configuration.

   b. Enzymes are configured (folded) so as to form grooves or 'pockets' which act as active sites. Therefore, any change in the three dimensional structure will alter the active site(s) and thereby affect binding of substrate to the enzyme.

Procedure (Quantitative)

p. 10-3   **Table 10A-1 Effect of Temperature on Enzyme Activity**

Minutes	Absorbance (420 nm)		
	at 10° C	at 24° C	at 50° C
0	0.09	0.05	0.06
2	0.25	0.43	0.07
4	0.28	0.58	0.08
6	0.39	0.65	0.08
8	0.45	0.68	0.09
10	0.46	0.68	0.08

Enzyme Activity at 6 Minutes

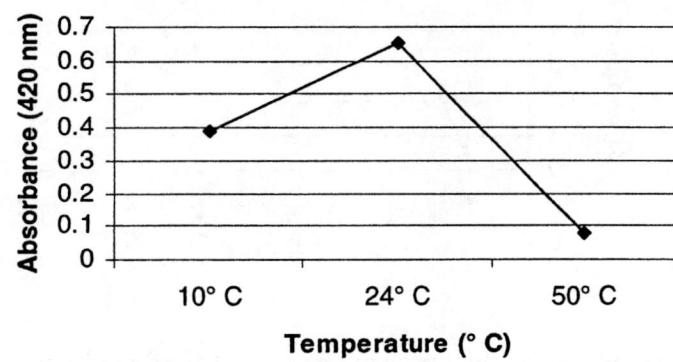

   c. 24° C.

p. 10-4   **d.** The rate can be measured as the amount of product produced per minute.

   e. 24° C

---

f. Yes, the results support the explanation at the beginning of the exercise. This enzyme appears to work better at lower temperatures and denatures quickly as temperature increases.

Procedure (Qualitative)

p. 10-4   Table 10A-2 Effect of Temperature on Enzyme Activity

Minutes	Intensity of Color		
	10° C	24° C	50° C
5	+	+++	++

p. 10-4 **c.** 24° C
d. Rate can be measured as product produced per minute.
e. Shaking the tubes mixes the contents and thus ensures that enzyme and substrate molecules contact each other.

p. 10-5 **f.** 24° C.
Cooking hint: A cooler temperature slows the rate of the enzyme-catalyzed reaction which produces the brown color.

PART 2   The Effect of pH on Enzyme Action

p. 10-5 **a.** Extremes of pH distort the three dimensional configuration of the enzyme molecule by altering the charges on the component amino acids, and thus change the manner in which these amino acids bond to determine the secondary and tertiary structure.

Procedure (Quantitative)

p. 10-6   Table 10A-3 Effect of pH on Enzyme Activity

pH 4	pH 6	pH 7	pH 8	pH 10
0.006	0.73	0.66	0.34	0.08

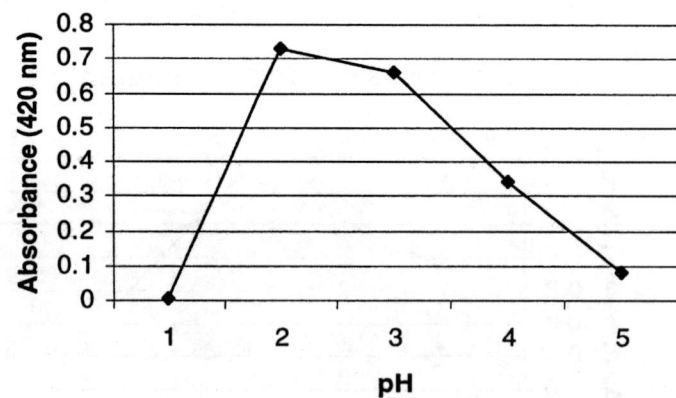

b. pH 6-7.
c. Rate is measured as the amount of product produced per minute.
d. pH 6-7.

p. 10-6   **Table 10A-4 Effect of pH on Enzyme Activity**

Intensity of Color				
pH 4	pH 6	pH 7	pH 8	pH 10
0	+++	+++	++	0

p. 10-7   **b.** pH 6-7.

          c. Rate is measured as the amount of product produced per minute.

          d. pH 6-7.

PART 3   **The Effect of Enzyme Concentration on Enzyme Activity**

p. 10-7   **a.** Increasing the substrate concentration will increase the rate of the enzyme-catalyzed reaction until sufficient substrate is added so that all of the enzyme molecules are continually involved in enzyme-substrate complexes. At this substrate concentration, the enzyme concentration becomes limiting and reaction rate no longer increases.

p. 10-8   **Table 10A-5 Effect of Enzyme Concentration on Enzyme-catalyzed Reactions**

	Absorbance (420 nm)			
Minutes	A	B	C	D
0	0	0.04	0.10	0.20
2	0	0.26	0.39	0.58
4	0	0.32	0.61	0.65
6	0	0.38	0.66	0.61

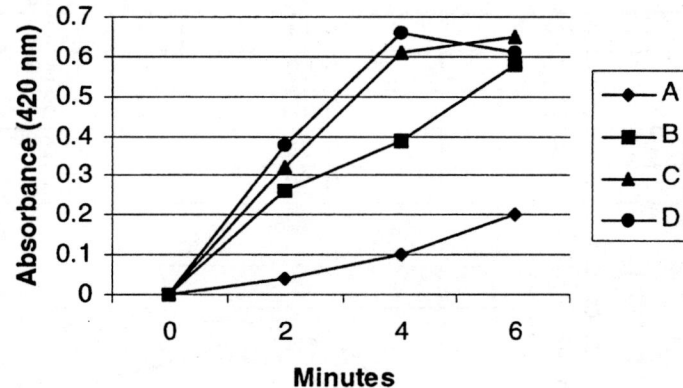

          b. The reaction rate is zero (no reaction) in tube A, which lacks enzyme. The reaction rate is fastest in tubes C and D, which contain the greatest concentrations of enzyme, but appears to be limited by the amount of enzyme as you reach 6 minutes.

p. 10-9   **c.** If substrate availability is unlimited, the reaction rate would continue to increase as enzyme concentration increased.

Procedure (Qualitative)

p. 10-9   **b.**   The brown color intensifies with time in tubes B, C, and D.

Table 10A-6 Effect of Enzyme Concentration on Enzyme-catalyzed Reactions

Intensity of Color			
A	**B**	**C**	**D**
0	+	+++	++

   c.   20 drops (color appeared faster in the D tube).
   d.   20 drops (color appeared faster in the D tube).

p. 10-10   **e.**   Yes. The reaction rate in tube C (2× enzyme) was approximately twice that in tube B (1× enzyme). The reaction rate in Tube D (4× enzyme) had the fastest rate but became substrate limited.

PART 4   The Effect of Substrate Concentration on Enzyme Activity

p. 10-10   **a.**   Since metabolic reactions are enzyme-catalyzed, the quantity of enzyme available and the intrinsic speed at which each enzyme can act will determine the rates of metabolic processes.

Procedure (Quantitative)

p. 10-11   **Table 10A-7 Effect of Substrate Concentration on Enzyme-catalyzed Reactions**

Tube	Absorbance (420 nm)			
	0 minutes	**2 minutes**	**4 minutes**	**6 minutes**
1	0	0.05	0.07	0.08
2	0.05	0.12	0.18	0.20
4	0	0.08	0.13	0.18
8	0.03	0.14	0.26	0.32
16	0.20	0.49	0.59	0.54
24	0.15	0.50	0.60	0.53
32	0.20	0.55	0.67	0.65
48	0.25	0.60	0.68	0.68

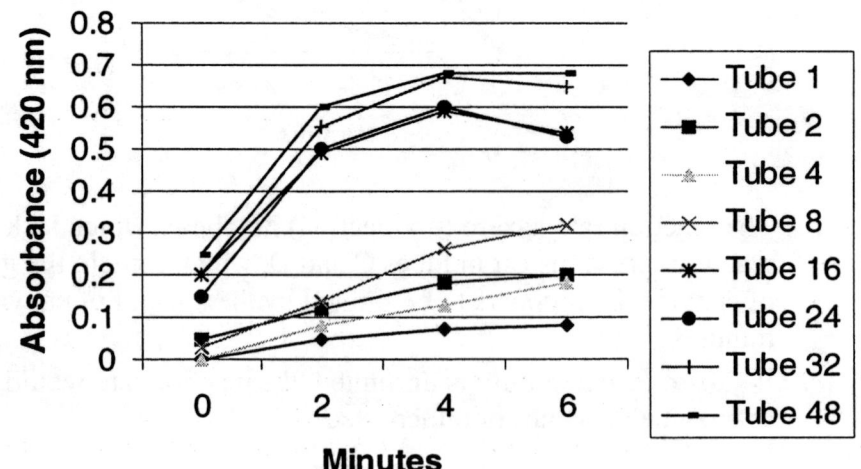

b. Yes. Reaction rate, measured as the change in absorbance, did not change when substrate concentration was increased from 16 to 24 drops.

c. The reaction rate would continue to increase.

p. 10-12 **d.** The curve would look like a straight line with an increasing slope.

Tube	Substrate Concentration	V_0
1	1 drop	0.025
2	2 drops	0.035
4	4 drops	0.040
8	8 drops	0.055
16	16 drops	0.145
24	24 drops	0.175
32	32 drops	0.175
48	48 drops	0.175

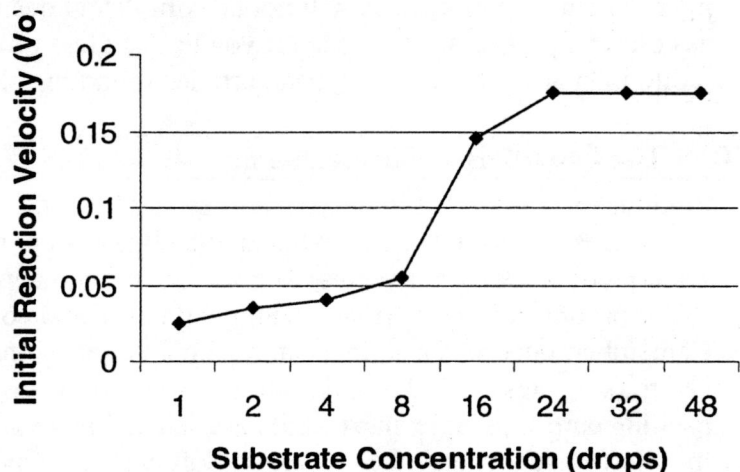

e. Yes. It appears that at a substrate concentration of 24 drops, the velocity at which the enzyme works is maximized.

12. For catecholase $V_{max} = 0.175$, $K_m = 9$ drops.

Procedure (Qualitative)

p. 10-14

Intensity of Color							
1	2	4	8	16	24	32	48
+	+	+	++	+++	+++	+++	+++

b. Yes. Increasing substrate concentration above 16 drops did not increase the velocity of the reaction because, at this substrate concentration, as enzyme molecules are continuously involved in enzyme substrate complexes.

Extending Your Investigation: Making Juices Juicier

p. 10-14   HYPOTHESIS: If pectinase is added to apple pulp (apple sauce), more juice can be extracted than if apple pulp remains untreated by the enzyme.

NULL HYPOTHESIS: The presence of enzymes makes no difference in the amount

---

of juice produced from fruits.

Prediction—twice as much juice can be collected if apple sauce is treated with pectinase.

Independent variable—amount of enzyme.

p. 10-15   Dependent Variable—amount of juice produced.

RESULTS:

Amount of juice collected without pectinase—4.6 ml.

Amount of juice collected with pectinase—7.7 ml.

Results support the hypothesis. Results allow the null hypothesis to be rejected

Conclusion—using pectinase increases the amount of juice that can be obtained from apple pulp (apple sauce).

Pectins are found in the middle lamella, a jelly-like layer that cements plant cells together. By disrupting this layer using pectinase, cells are freed from large clumps and cell membranes can be more easily ruptured to expel more fluid or juice. The pectin molecules will not be completely destroyed. Rather, they remain as colloidal particles, easily seen if you hold a glass of apple juice up to the light (colloidal particles are larger than particles found in solution, usually 1-1000 nm).

EXERCISE B   The Essentials of Cheesemaking

p. 10-16   **a.** Renin is most effective in making cheese at 25-30° C.

b. Blue cheese: this cheese is incubated with *Penicillium roquefortii* (fungi imperfectii or Deuteromycetes) to give the marbled appearance and tart taste of roquefort or blue cheese (made with goat's milk and cow's milk respectively). Camembert cheese is also incubated with a fungus (fungi imperfectii or Deuteromycetes) called *Penicillium camembertii*. Hard cheeses are made by packing curd into large slabs which are stacked in a vat and turned every 10 minutes, a process called cheddaring. Soft cheeses may have emulsifiers added. Herbs, seeds, alcohol, and vegetable dyes may also be incorporated.

c. Whey powder can be condensed and dried and is used in animal feeds and as a protein supplement for humans eating low protein diets. It is also used in the baking industry. The whey complex contains lactalbumin, lactoglobulin, and lactomucin.

Laboratory Review Questions and Problems

1.   Enzymes function optimally under specific conditions of pH and temperature (in this experiment, optimum conditions were pH = 6-7, Temperature = 24° C). Increases in enzyme or substrate concentration will increase reaction rate. However, rate increases will be limited by substrate concentration or by enzyme concentration, unless both enzyme and substrate are available without limit. The reaction rate will also be limited by the intrinsic maximum velocity by which the enzyme itself acts. This is dependent on the attraction (affinity) of enzymes for substrate.

2. Because enzymes will function only in defined environments, the activity of enzymes (and therefore the processes of metabolism) can be controlled by alterations in the environment. The rates of enzyme-catalyzed reactions can also be controlled by increased or decreased

synthesis of the enzymes, since greater concentrations of enzyme increase reaction rate. The supply of substrate will also limit the rate of enzyme-catalyzed reactions.

3. When proteins are denatured, their secondary and tertiary structures are destroyed because hydrogen bonds and disulfide bonds holding parts of protein chains together (amino acids form hydrogen bonds or disulfide bonds with other amino acids) are broken. Heat and changes in pH can denature enzymes.

4. If 50 molecules of enzyme are present but only 25 molecules of substrate are present, then the enzyme cannot act at its maximum rate. If all molecules of enzyme could be engaged with substrate, much more product would be produced per unit time. Thus, the enzymatic reaction is limited in its rate by the lack of substrate (substrate-limited).

5. 9 drops of catecholase.

6. Enzymes with a high affinity for the substrate have a low K_m. In the graph below, the reaction A is much slower (gentler slope) than reaction B. The K_m is much higher for the enzyme catalyzing reaction A because it has less affinity for the substrate; hence the slower reaction rate. The enzyme catalyzing reaction B has a much smaller K_m and a higher affinity for the substrate; hence a faster reaction rate.

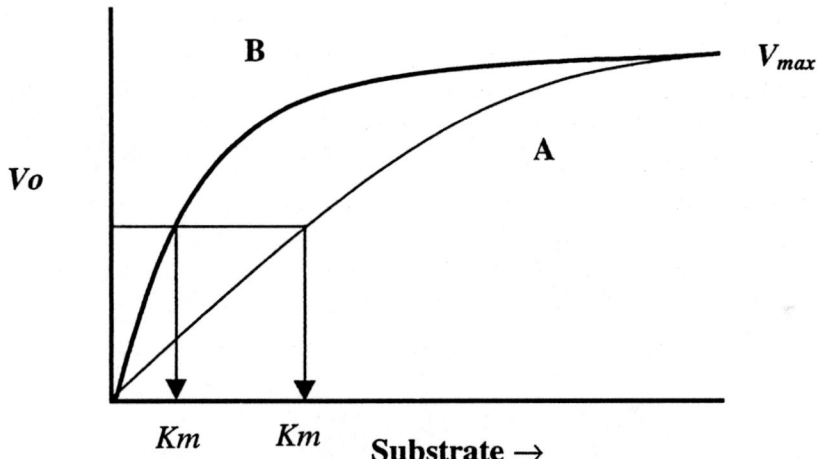

7. (a) K_m is higher for the enzyme when a competitive inhibitor is present. For example, if 50 molecules of enzyme are present and 50 of substrate plus 50 of inhibitor, it is just as likely that an inhibitor molecule will bind with the enzyme instead of the actual substrate. It will take longer for the enzymes to "find" the proper substrate molecules among the diverse mixture of substrate and inhibitor. (b) However, as you increase substrate, there is more and more of a chance that the substrate rather than the inhibitor will bind. With enough substrate present, you can actually "out-compete" the competitive inhibitor. This is the reason that you see the curves for both reactions eventually reaching the same V_{max}. (c) The reaction rate of the non-inhibited reaction is faster than the inhibited reaction until the effects of inhibitor are completely overshadowed by the presence of enough substrate. At this point, both reactions have the same V_{max}.

8. (a) The K_m for an enzyme inhibited by a non-competitive inhibitor does not change. Usually

noncompetitive inhibitors bind to sites on the enzyme other than the active site, or they are very small molecules (for example, mercury) which can bind to the active site but still allow the substrate to bind. The enzyme still works, but a little more slowly (like walking is slowed when you have a pebble in your shoe). (b) For this reason, the V_{max} will always be different; lower for the inhibited reaction. (c) The enzyme binds just as much substrate whether it is inhibited or not, but it just cannot work as fast when the inhibitor is bound. Even adding more substrate does *not* "out-compete" the competitive inhibitor in this case because the enzyme molecules are already bound to substrate and addition of more substrate does not increase the rate of the reaction.

Laboratory 11

Energetics, Fermentation, and Respiration

EXERCISE A   Production of Carbon Dioxide and Ethanol by Fermentation

PART 1   Examining Yeast Cells

p. 11-3   a. Neutral red stains nuclei red, to allow them to be seen.
          b. Yeast cells are eukaryotic because they have nuclei.
          c. Glycolysis takes place in the cytoplasm.

PART 2   Production of Carbon Dioxide by Fermentation

p. 11-3   a. Carbon dioxide is evolved from the flask containing sugar and yeast.
          b. Carbon dioxide is produced during fermentation by the yeast cells.
          c. Bubbles are formed by carbon dioxide being released.

PART 3   Requirements for Fermentation in Yeast

p.11-4   HYPOTHESIS: If glucose is the substrate with which glycolysis begins, then it will
          be used more quickly and result in a higher rate of fermentation.

          NULL HYPOTHESIS: All sugars will support the process of glycolysis and
          fermentation at equal rates.

          Prediction—Group I will exhibit the highest rate of photosynthesis. Group II will
          show the second highest rate. Fermentation should not occur in Group III with
          boiled (dead) yeast nor in Group IV without sucrose (substrate).

          Independent variable—time.

          Dependent variable—amount of CO_2 produced.

p. 11-5   **Table 11A-2 Class Data for Yeast Fermentation Experiment, Recorded as
          Length of Gas Column (in millimeters)**

	Group I Student Pair					Group II Student Pair					Group III Student Pair					Group IV Student Pair				
Min	1	2	2	4	Avg	1	2	3	4	Avg	1	2	3	4	Avg	1	2	3	4	Avg
0	0	0	0	0	0	0	0	0	0	0	0	0	0	0	0	0	0	0	0	0
5	1	0	0	0	0.25	0	0	0	0	0	0	0	0	0	0	0	0	0	0	0
10	3	1	0.5	0.5	1.25	0	0	0	0	0	0	0	0	0	0	0	0	0	0	0
15	3	2	1	1	1.75	0.3	0.2	0.5	0.2	0.3	0	0	0	0	0	0	0	0	0	0
20	5	3	1	2	2.75	0.4	0.2	0.5	0.2	0.3	0	0	0	0	0	0	0	0	0	0
25	5	3	1.5	2	2.89	0.5	0.3	1	0.2	0.5	0	0	0	0	0	0	0	0	0	0
30	5	3	1.5	2.5	3	0.5	0.4	1	0.2	0.5	0	0	0	0	0	0	0	0	0	0
35	5	3	1.5	2.5	3	0.8	0.5	1.5	0.2	0.9	0	0	0	0	0	0	0	0	0	0
40	5	3	1.5	2.5	3	1	0.6	1.5	0.5	0.9	0	0	0	0	0	0	0	0	0	0
45	5	3	1.5	2.5	3	1.5	1	2	0.7	1.3	0	0	0	0	0	0	0	0	0	0

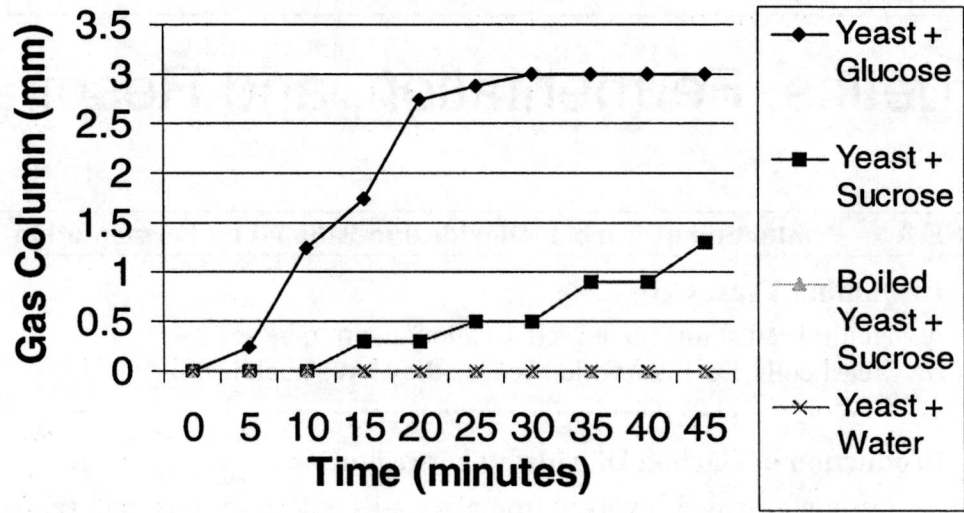

11. Treatment Group I

$$\frac{1.25 - 0.25}{10 - 5} = \frac{1.01}{5} = 0.2 mm/\min$$

$V = \pi r^2 (h) = \pi (18^2) 0.2 = 203.6 \text{ mm}^3$

Treatment Group II

$$\frac{0.5 - 0.3}{25 - 15} = \frac{0.2}{10} = 0.02 mm/\min$$

$V = \pi r^2 (h) = \pi (18^2) 0.02 = 20.4 \text{ mm}^3$

Treatment Groups III and IV

$$0.0 mm/\min$$

Table 11A-3

Treatment Group	Fermentation Rate (ml/min)
I 5% glucose	
II 5% sucrose	
III boiled yeast	0
IV no sucrose	0

a. *Group I.* Yeast utilize glucose for anaerobic respiration and fermentation.
   Group II. Yeast utilize sucrose for anaerobic respiration and fermentation at a slower rate than they utilize glucose.
   Group III. Boiled yeast do not ferment substrate.
   Group IV. Yeast need an energy source (e.g., glucose) for fermentation to occur.
b. Group IV is a control. It is used to show that something else in the mixture (other than sugar) could not be responsible for fermentation results.
c. Boiling yeast denatures the enzymes responsible for glycolysis and fermentation.
d. Group III is also a control for this experiment. It is used to show that yeast, and nothing else in the reaction mixture, is responsible for fermentation. You could add assays of carbon dioxide production by sucrose without yeast, and glucose without yeast, to ensure that the sugars themselves are not sources of carbon dioxide without the action of yeasts.

Results support the hypothesis.

Results allow the null hypothesis to be rejected.

Conclusion—yeasts carry out the process of anaerobic respiration and fermentation. Glucose is used more quickly, but other sugars that can be broken down or isomerized to form glucose can also be used (but at a slower rate).

PART 4   Production of Ethanol by Yeast

p. 11-7   **a.** Alcohol is formed by anaerobic fermentation by the yeast cells.

EXERCISE B   The Krebs Cycle Reactions in Bean Seeds

p. 11-9   **Table 11B-2 Absorbance of Bean Seedling Solutions**

Minutes	Tube 1	Tube 2	Tube 3	Tube 4	Tube 5	Tube 6
0	0.65	0.70	0.70	0.79	0.74	0.93
2	0.52	0.69	0.69	0.82	0.67	0.94
4	0.35	0.68	0.68	0.80	0.62	0.92
6	0.36	0.68	0.68	0.81	0.56	0.90
8	0.33	0.67	0.68	0.80	0.46	0.89
10	0.25	0.67	0.67	0.79	0.37	0.82
12	0.19	0.66	0.67	0.78	0.30	0.84
14	0.17	0.66	0.66	0.79	0.24	0.82
16	0.11	0.66	0.66	0.79	0.018	0.79

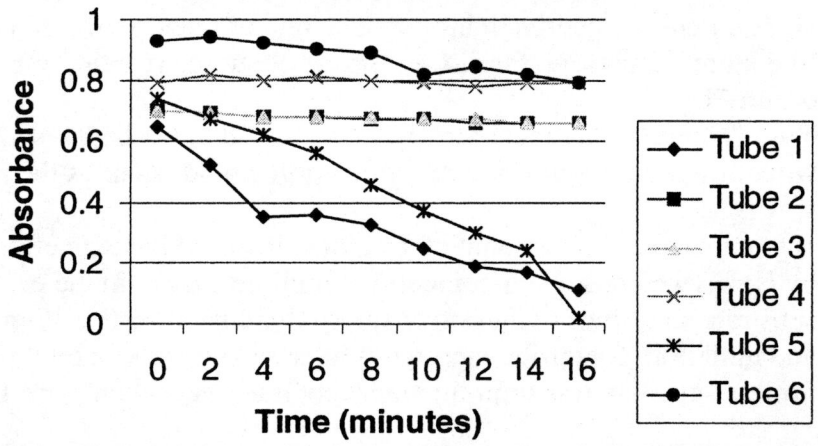

Bean Seedling Solutions

9. <u>Tube 1</u>: Contained all the necessary components. Hydrogen electrons that would have been used in the synthesis of fumarate were instead used to decolorize DPIP.

<u>Tube 2</u>: Lacked substrate. No hydrogen electrons were produced, therefore DPIP was not decolorized.

<u>Tube 3</u>: Enzymes were denatured by boiling. No hydrogen electrons were produced, therefore DPIP was not decolorized.

<u>Tube 4</u>: The inhibitor (malonate) stopped the reaction. No electrons were produced, therefore DPIP was not decolorized.

Tube 5: Reaction rate was increased by adding twice the substrate concentration. Rate will continue to increase until the reaction is limited by enzyme concentration.

Tube 6: The inhibitor (malonate) was not as effective in stopping the reaction when three times the amount of substrate (succinate) was present. (This is because malonate is a competitive inhibitor and has been out-competed by excess substrate.)

a. Production of electrons and decolorization of DPIP indicates that the beans contained functional enzymes, and therefore were alive.

p. 11-10 **b.** When lima bean juice was boiled, the enzymes were denatured. Therefore, no enzymes were present to break down substrate and release hydrogen electrons so DPIP was not decolorized.

b. Tube 2 is a control. It lacked substrate.

d. In tube 4, the reaction was inhibited by malonate.

e. Malonate is a competitive inhibitor and when it is included in the reaction mixture with succinate, the enzyme may encounter malonate instead of succinate. This would reduce the amount of product produced per unit of time, thus reducing the rate of the enzymatic reaction.

f. More fumarate is produced in tube 6. More decolorization of DPIP occurs and this only happens when succinate is converted to fumarate, releasing hydrogen ions that can reduce DPIP.

p. 11-11 **g.** Tube 6 contains more succinate.

h. When the amount of succinate is increased it out-competes malonate. In other words, the chances of an enzyme encountering succinate rather than malonate are increased the more succinate is present in the reaction mixture.

i. Hg is a non-competitive inhibitor. The rate of reaction would be greatly slowed (the more Hg present, the slower the rate), but the reaction would still have occurred.

j. Yes. The maximum rate of the reaction would never be the same as the reaction without inhibitor, but some decolorization would occur as the reaction proceeded slowly.

k. If you increased the amount of succinate it would have no effect on the inhibition by $HgCl_2$ which is a non-competitive inhibitor. Even in the presence of $HgCl_2$, substrate still binds to the active site so there is no competition between substrate and inhibitor. Therefore, increasing substrate concentration will have no effect (unless substrate was limiting the reaction to begin with—see Laboratory 10).

Extending Your Investigation: Studying Inhibition

p. 11-11 HYPOTHESIS: If malonate is a competitive inhibitor, then increasing amounts of succinate would eventually allow you to out-compete the action of the inhibitor, malonate.

NULL HYPOTHESIS: The presence of increasing amounts of succinate will have no effect on reaction rate.

Prediction—as the succinate/malonate ratio increases, more decolorization of DPIP will take place.

Independent variable—time.

Dependent variable—color of DPIP-containing solutions.

p. 11-12 PROCEDURE: Same as Exercise B, using the following:

Minutes	Tube 1	Tube 2	Tube 3	Tube 4	Tube 5	Tube 6
0	0.70	0.80	0.89	0.83	0.86	0.62
2	0.67	0.75	0.63	0.84	0.83	0.63
4	0.62	0.70	0.60	0.82	0.78	0.62
6	0.57	0.68	0.61	0.81	0.73	0.62
8	0.50	0.58	0.60	0.80	0.67	0.58
10	0.43	0.53	0.62	0.79	0.64	0.58
12	0.41	0.47	0.61	0.78	0.59	0.56
14	0.36	0.43	0.60	0.78	0.54	0.55
16	0.32	0.37	0.60	0.76	0.52	0.55

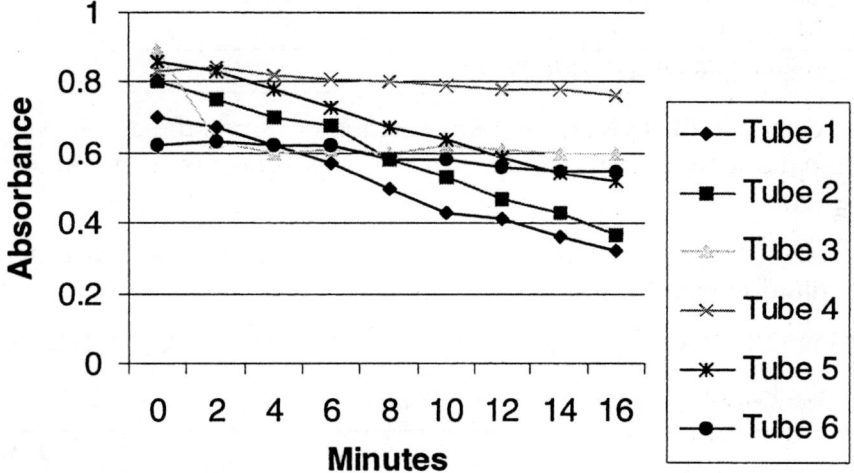

Results support the hypothesis. Increasing the amount of succinate causes the reaction rate to increase.

Results allow the null hypothesis to be rejected.

Prediction—correct. As the amount of succinate increased, the amount of decolorization was also increased.

Conclusion—competitive inhibitors can be out-competed by increasing substrate to a higher level, in this experiment about an 8-fold excess.

EXERCISE C   Heat Production During Respiration in Seedlings

p. 11-13

Time (hours)	Thermos Temperature (° C)	Room Temperature (° C)
5	18.5	18
10	19	18
25	21	17.5
30	22	18
50	23	18

a. Yes. The increase in temperature indicated that heat was produced in respiration.

---

    b. Cotyledon endosperm.

    c. Small increases in temperature would increase respiration rate. High temperatures would eventually denature the respiratory enzymes.

    d. The rate of enzyme-catalyzed reactions approximately doubles with each 10°C rise in temperature, but very high temperatures denature enzymes.

    e. Metabolic heat can be used to maintain constant body temperature.

p. 11-13   **f.** 63% efficiency.

EXERCISE D   Respiration in Plant Embryos

p. 11-14   **a.** The embryo.

p. 11-15   **b.** The group in which the embryos stained red was alive.

    c. Cyanide treated seeds would not stain red. Cyanide inhibits electron transfer. Since electrons are needed to produce a red color with TTC, cyanide-treated seeds would not stain red.

Laboratory Review Questions and Problems

1. Glycolysis occurs in the cytoplasm. Fermentation occurs in the cytoplasm. Krebs cycle occurs in the mitochondrial matrix. Electron transport occurs in the mitochondrial inner membrane (cristae).

2. Fermentation is an inefficient process because it generates only 2 ATP/glucose, whereas aerobic respiration generates 36-38 ATP/glucose.

3. Fermentation is an anaerobic process because it does not use O_2 to degrade glucose.

4. Requirements Products Final Electron Acceptor

	Requirements	Products	Final Electron Acceptor
Alcoholic fermentation	pyruvate	CO_2, ethanol, ATP	ethanol, no ETS
Lactate fermentation	pyruvate	lactic acid	lactic acid, no ETS
Aerobic respiration	glucose, O_2	CO_2, ATP, H_2O	O2

5. Heavy breathing can restore oxygen levels in the blood. In muscle cells that have become depleted of oxygen, cells have respired anaerobically to produce lactic acid and deplete reserves of creatine phosphate. Under aerobic conditions, lactic acid can be converted to pyruvate and creatine phosphate reserves can be regenerated. Soreness and cramping will gradually disappear.

6. Solution A: inhibitor
   Solution B: substrate
   Solution C: enzyme
   The inhibitor is a competitive inhibitor based on the results from tubes 2 and 7.

7. Twenty (20) pyruvate molecules; 20 acetyl molecules; 40 carbons. Although fatty acids are usually not this long (two 20 carbon fatty acids are possible). Example of fatty acid with only 10 carbons:

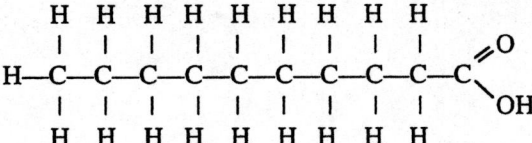

8. To test for the percentage of seeds likely to germinate, the farmer could do the following: 1) Test a sample of the seeds with TTC. Viable seeds will stain red. 2) Calculate the percentage of viable seeds in the sample. This is the percentage of viable seeds in the entire bag of seeds.

Laboratory 12

Photosynthesis

EXERCISE A   The Energy-Capturing Reactions

PART 1   Determining the Effect of Light Intensity

p. 12-2   HYPOTHESIS: If light affects photosynthesis then the brighter (more intense) the light, the more photosynthesis (higher rate) will occur.

NULL HYPOTHESIS: No difference in the amount or rate of photosynthesis will be seen with different light intensities.

Prediction—More disks will float in Petri dishes exposed to higher light intensities than in lower light intensities.

Independent variable—amount of light.

Dependent variable—number of disks floating.

p. 12-4   Table 12A-2 Effect of Varying Light Intensities on Photosynthesis, Measured as Number of Floating Disks

Intensity (W/m2)	Group						Mean
	1	2	3	4	5	6	
0.000 00 W	3	2	0	5.6	3	2	3
0.036 40 W	8	16	7	10	11	9	10
0.064 60 W	9	10	13	10	17	8	11
0.100 100 W	13	11	19	12	17	13	17
0.154 150 W	12	20	18	19	20	19	18

11. % floating discs

Light Intensity (W/m^2)	% of Disks Floating
0.000	15
0.036	50
0.064	55
0.100	70
0.154	90

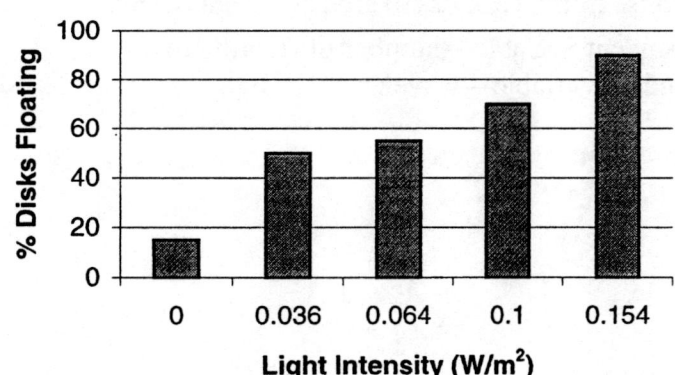

a. It is important to have as many replicates of the experiment as possible. Means provide a way of presenting the "best possible" representation of the array of data. If only one trial is used, something might have affected the results. For instance, disks may not have been degassed completely or may be exposed to light while being transferred into the experimental dishes. Note that even in 0 light, some disks floated indicating the latter two possibilities may have actually occurred. Results support the hypothesis.

Results allow the null hypothesis to be rejected.

Conclusion—the greater the light intensity used to drive photosynthesis, the faster the photosynthetic rate and the more CO_2 is evolved during the process.

The table below represents class data from a week of experiments on the average, maximum, and minimum number of floating spinach disks in 0.2% sodium bicarbonate after 10 minutes of exposure to varying intensities of light. Chi-square median probability values are compiled using the statistical package on the BioBytes CD-ROM.

Bulb (watts)	Intensity (W/m2)	Mean	Maximum	Minimum
0	0.000	2.212	5.6	0.3
40	0.036	10.049	18.0	5.4
60	0.064	13.099	17.5	9.5
100	0.100	15.512	19.0	12.3
150	0.154	16.137	18.3	11.0

Chi-square = 10.0, Probability = 1.0 - 2.5%

With 5 treatments, the degrees of freedom = 4. With a chi-square median value of 10, this is between a 1 in 20 and 1 in 100 chance that the results are due to chance alone. This means that the data are statistically significant and the null hypothesis can be rejected. (Note: The chi-square median test was used as a comparison of values with the mean rather than using a typical chi-square test that evaluates numerical counts data.)

PART 2 The Spectrum of Visible Light

p. 12-6   HYPOTHESIS: If photosynthesis requires absorption of photons of light then blue and red wavelengths of light will support increased photosynthesis.

NULL HYPOTHESIS: All wavelengths of light support the photosynthetic process equally.

Prediction—More spinach disks will float in dishes covered by red and blue filters than those in the dark or covered by green filters.

p. 12-7   Independent variable—number of floating disks.

Dependent variable—wavelength of light.

Table 12A-4 Effects of Different Wavelengths of Light on Photosynthesis, Measured as Number of Floating Disks

Wavelength (λ)	Group						Mean
	1	2	3	4	5	6	
White (380-470 nm)	18	20	16	18	17	19	18.0
Blue (430-490 nm)	12	11	7	12	10	13	10.8
Green	7	5	8	3	7	6	6.0
Red	13	11	8	7	10	9	9.7
Dark (no light)	2	1	3	2	1	3	2.0

11. % floating disks...

Wavelength	% Disks floating
White	90
Blue	54
Green	30
Red	48.5
Dark	10

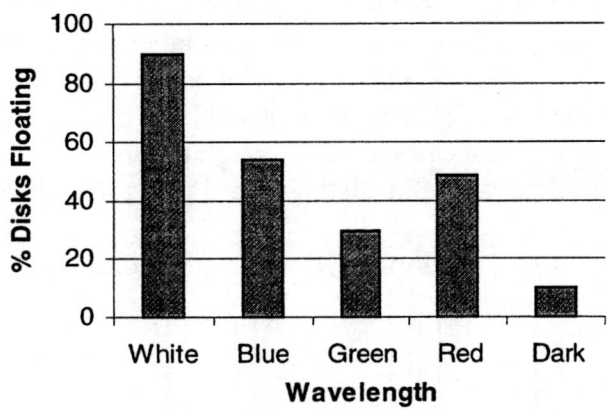

a. In this experiment, we have measured the *process* of photosynthesis—its action as a result of absoprtion of different wavelengths of light by all pigments in the leaf. This can be compared to an absoprtion spectrum (see Exercise B, Part 2) which determines what wavelengths of light are absorbed by each pigment.

p. 12-8   Results support the hypothesis.
Results allow the null hypothesis to be rejected.

b. Red (610-700 nm) and blue (450-500 nm) wavelengths are most effective in promoting photosynthesis.

c. Green (500-570 nm) light is least effective in promoting photosynthesis. It is not absorbed by plants but is reflected which is why green plants appear to be green. Conclusion—different wavelengths of light affect the photosynthetic process. Red and blue light support the highest rates of photosynthesis (measured as amount of CO_2 produced/10 minutes as demonstrated by the ability of disks to float).

The data below represent chi-square median test data for class results over a weeklong period of experiments on the number of spinach disks floating when exposed to varying wavelengths of light.

Filter color	Mean	Maximum	Minimum
Blue	9.3	12	5
Green	4.1	10.5	0.3
Red	10.8	15.5	7

Chi square = 7, Probability = 1 – 2.5%

With 3 treatments, the degrees of freedom (df) = 2. With a chi-square median value of 7, there is between a 1 in 20 chance and a 1 in 100 chance that the results are due to chance alone. This means that the data are statistically significant and the null hypothesis can be rejected.

EXERCISE B   The Pigments of Chloroplasts

PART 1   How Plant Pigments Use the Light Spectrum

p. 12-10
   a. Carotenes are the most nonpolar pigments.
   b. Chlorophyll b is the most polar pigment.
   c. Chlorophylls a and b contain Mg^{2+} associated with a tetrapyrrole ring. This charged ion causes the molecules to be polar.
   d. Two pigments were chlorophylls (a and b).
   e. Two pigments were carotenoids (though plants may contain a variety of carotenoids that cannot be distinguished using this chromatographic technique).
   7.

chlorophyll a   chlorophyll b

carotenes   xanthophyhlls
xanthophyll breakdown
products

Note: R_f values can be calculated for each band. The R_f expresses the relationship between the distance moved by the solvent and distance moved by the pigment.

$$R_f = \frac{\text{distance pigment migrated (mm)}}{\text{distance solvent front migrated (mm)}}$$

Band number	Distance (mm)	Band Color
1	16	yellow-green
2	31	blue-green
3	54	yellow
4	62	orange/yellow

R_f = 0.97, carotene (yellow to yellow orange)
R_f = 0.56, xanthophyll (yellow)
R_f = 0.32, chlorophyll a (bright green to blue green)
R_f = 0.16, chlorophyll b (yellow green to olive green)

PART 2 Absorption Spectra of Chloroplast Pigments

p. 12-10,11 1. At 450 nm you see red. At 550 nm you see green. At 650 nm, you see blue.

p. 12-11 **Figure 12B-2** *Absorption spectra of chloroplast pigments*

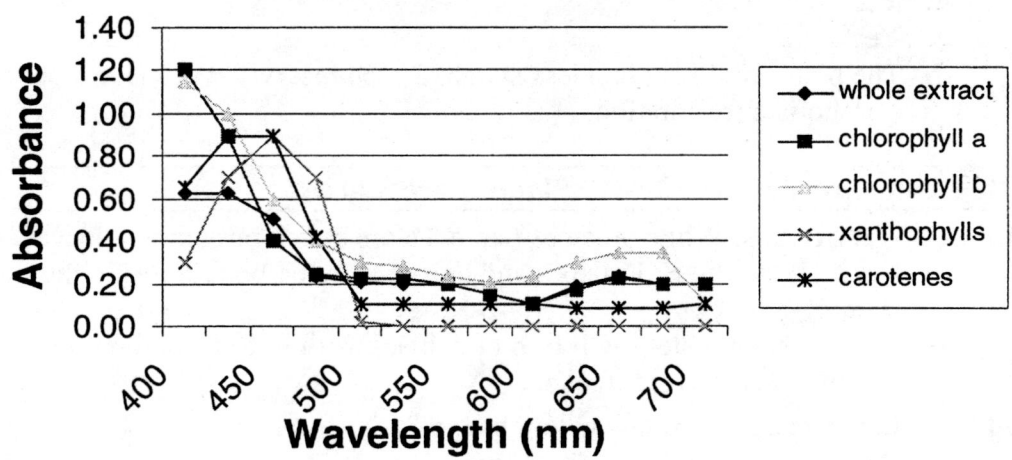

a. An action spectrum tells which wavelengths of light support the photosynthetic process. An absorption spectrum tells you which pigment absorbs what wavelengths of light. By comparing the two, you can determine which wavelengths of light are absorbed by pigments to drive the photosynthetic process.

p. 12-12 a. Chlorophyll a is the primary photosynthetic pigment. Chlorophyll b and the carotenoids are accessory photosynthetic pigments, which harvest light energy for use by chlorophyll a. The absorption spectrum for the combined chlorophylls and carotenoids matches the action spectrum for photosynthesis, thus all of the pigments are responsible for photosynthesis.

PART 3 The Light-Independent Reactions of Photosynthesis

p. 12-12 a. Leaves of light grown seedlings are robust, green and healthy.
b. Leaves of dark grown seedlings are yellow to white (etiolated). The plants are spindly.
c. Chlorophyll is present in light grown plants and is not present in dark grown plants.

d. Yes

e. Yes. Spinach disks placed in the dark did not float.

f. Dark grown plants can survive and grow for a while, using the energy reserves (sugars) stored in the endosperm of the cotyledons of the seeds.

EXERCISE C   The Light-Independent Reactions of Photosynthesis

p. 12-10   Table 12C-1 Relation of Plant Pigments to Starch Formation

Pigment	Starch Present?
none	no
anthocyanins	no
carotenes/xanthophylls	no

a. Carbon-fixing reactions resulted in starch formation in the green (chlorophyll-containing) parts of the plants.

b. Yes

c. Yes

d. The plant would contain less starch. Starch reserves would have been used in metabolism (respiration).

Extending Your Investigation: Do All Plants Store Starch?

p. 12-13 HYPOTHESIS: White potatoes (stems) store starch but onions (leaves) store sugar.

NULL HYPOTHESIS: Potatoes and onions store carbohydrates in the same way (as starch).

Prediction—potatoes will turn blue-black with Lugol's and onions will form a red precipitate with Lugol's.

p. 12-14 Independent variable—color change.

Dependent variable—reagent type.

a. The potato turns blue-black, but the onion remains brownish.

b. The potato contains starch.

c. The tube that does not contain plant material serves as a control.

RESULTS:

Material	Color at Start	Color After 3 Minutes	Presence of Precipitate
.onion	blue	red	yes
potato	blue	blue	no
no material	blue	blue	no

Discussion—the onion material caused Benedict's to turn red so it must have contained sugar, but the potato did not produce a red precipitate so it does not contain free glucose (prolonged boiling may produce an orangish precipitate as starch is broken down to form dextrins and eventually glucose).

Results support the hypothesis.

Results allow the null hypothesis to be rejected.

Conclusion—onions store carbohydrates as sugar while potatoes store starch.

p. 12-15 Do potatoes store starch? Yes

Do potatoes store sugar? No

Do onions store starch? No
Do onions store sugar? Yes

Plants	Type of sugar
Idaho potato	starch
purple onion	glucose

Laboratory Questions and Problems

1. **a.** Light as the energy source.
   b. CO_2 as the carbon source
   c. Chlorophyll (and, potentially, accessory pigments) for the absorption of light energy.
   d. Water as the electron donor.

2. Red and Blue light are used in photosynthesis by vascular plants.

3. **a.** An action spectrum indicates the wavelengths of light at which photosynthesis occurs and is most efficient. An absorbance spectrum indicates the aborbance of various wavelengths of light by the pigments of a plant.
   b. The action spectrum matches the absorbance spectra of chlorophylls a and b at wavelengths of 650-700 nm, and the carotenoids at wavelengths of 450-500 nm.
   c. Plants contain many pigments to maximize utilization of sunlight (white light) which contains all wavelengths (colors). Chlorophyll a is the primary photosynthetic pigment. Chlorophyll b and the carotenoids (carotenes and xanthophylls) absorb excess light and pass the energy obtained to chlorophyll a molecules. Additionally, carotenoids may serve as photoprotectants, protecting the chlorophyll from intense, potentially damaging, light.

4. The colored pigments might mask the color produced by the I_2KI.

5. Anthocyanins are water-soluble, polar molecules and therefore do not extract in the nonpolar solvents used to prepare the chlorophyll extract.

6. **a.** The compensation point is reached at 35° C.
   b. Growth would be most rapid at 20° C. At temperatures up to 35° C, the rate of photosynthesis exceeds the rate of respiration. Since growth is dependent upon the supply of energy (as photosynthate, glucose) available in excess of the cell's basic metabolic needs, growth is most rapid at the temperature where there is the greatest difference between glucose synthesis (photosynthesis) and glucose utilization (respiration). At 20° C, the rate of photosynthesis exceeds the rate of respiration by a factor of three. Therefore, growth would be most rapid at 20° C.
   c. As temperature increases, rates of photosynthesis and respiration increase, until the temperature exceeds the optimum temperatures for the enzymes catalyzing reactions. Both photosynthetic and respiratory rates decrease at high temperatures (>45° C). Additionally, photosynthetic rates may decrease at moderately high temperatures (>28° C in the example) due to CO_2 limitation as leaves close their stomates.
   d. Cool temperatures and a carbon dioxide-rich atmosphere minimize respiration rates. Therefore, fruits stored under these conditions maintain their sugar reserves for longer periods of time.

7. CO_2 concentration could limit the rate of photosynthesis because CO_2 is necessary for the synthesis reactions of the Calvin Cycle: Low CO_2 concentration results in low photosynthetic rate. Light quantity could limit the rate of photosynthesis by limiting the number of photons of light energy passed into the electron transport system, and thus limiting the number of ATPs produced. Light quality could limit the rate of photosynthesis. For a plant that was illuminated with light of a color (wavelength) that was not absorbed by the photosynthetic pigments of that plant, the light energy would not be harvested and the photosynthetic rate would be diminished. For example, vascular plants typically cannot use green light for photosynthesis, since they lack pigments that absorb light of 500-570 nm. Temperature increases photosynthetic rate to an optimum temperature, then photosynthetic rate decreases as CO_2 supplies diminish when the plants close their stomates. Water is necessary as the electron donor in the light-capturing reactions of photosynthesis.

Laboratory 13

Meiosis: Independent Assortment and Segregation

EXERCISE A   Simulation of Chromosomal Events During Meiosis

p. 13-4 **a.** No.

b. In metaphase of mitosis, dyad chromosomes (each made up of two sister chromatids) line up in single file on the metaphase plate. In meiosis, homologous pairs (tetrads) line up (i.e., chromosomes are arranged in pairs rather than single file).

p. 13-5 **c.** In meiosis, each daughter cell produced by Meiosis I receives one intact dyad (one chromosome with 2 chromatids). In mitosis, each daughter cells receives a single chromatid of each dyad.

d. Chromatids do not separate; they remain attached at the centromere region. Therefore, each chromosome consists of two chromatids at anaphase of Meiosis I.

e. The amount of DNA is half that of the cells entering meiosis. Recall that the $2n$ cell duplicates its DNA (to a $4n$ amount during interphase). The cells are haploid at the end of Telophase I (they have one-half the number of chromosomes) but each chromosome still has two chromatids (so the total amount of DNA is still a $2n$ amount).

f. One of each *type* of chromosome is in each cell but each chromosome has two chromatids.

p. 13-6 **g.** Yes

h. In prophase I, homologous chromosomes synapse to form tetrads. No synapsis occurs in Prophase II.

p. 13-7 **i.** Individual chromosomes (rather than homologous pairs) line up on the metaphase plate in Metaphase II.

j. In both Metaphase II of meiosis and Metaphase of mitosis, individual double-stranded (dyed) chromosomes line up in single file.

p. 13-8 **k.** 4

l. 2

m. $1n$ cells formed by meiosis; $2n$ cells formed by mitosis.

n. $2n$ (diploid).

o. (1) Homologous chromosomes synapse during Prophase I of meiosis; (2) Homologous pairs line up during metaphase I of meiosis; (3) DNA replication does not occur during Interphase II (interkinesis) of meiosis. The final end products are different; mitosis produces two $2n$ cells whereas meiosis produces four $1n$ cells.

Yes, haploid, yes, no.

Yes, diploid, yes, yes.

p. 13-8   **a.** Because the chromatids are copies of one another, so, if *A* is on one chromatid, then *A* must also be on the other chromatid.

p. 13-10   Products of meiosis:

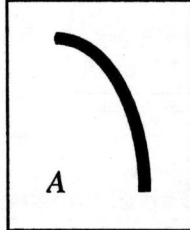 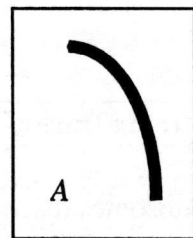 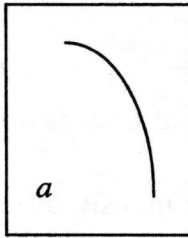 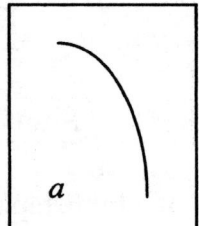

b. Alleles segregated into individual cells. Each 1*n* cell contains one copy of one of the homologues.

c. Yes. Each allele is segregated into an individual cell.

p. 13-11   **d.** Purple

e. *R*

f. Yellow

g. *r*

h. *Rr*

i. Purple

j. *R, r*

	R	*r*
R	*RR*	*Rr*
r	*Rr*	*rr*

Genotype	Phenotype
RR	purple
Rr	purple
Rr	purple
rr	yellow

k. 3.

l. *RR*, 1: *Rr*, 2: *rr*, 1

m. 2

p. 13-12   **n.** 3 purple: 1 yellow

o. Good comparison (3.2 = experimental; 3 = predicted)

p. No. Cannot determine whether the purple kernels are *RR* or *Rr*.

p. 13-12   **a.** Since alleles of unlinked genes assort independently, any of the possible genotypes may be present in a gamete.

p. 13-14, 15   **Figure 13C-3** *Meiosis with two pairs of homologous chromosomes*

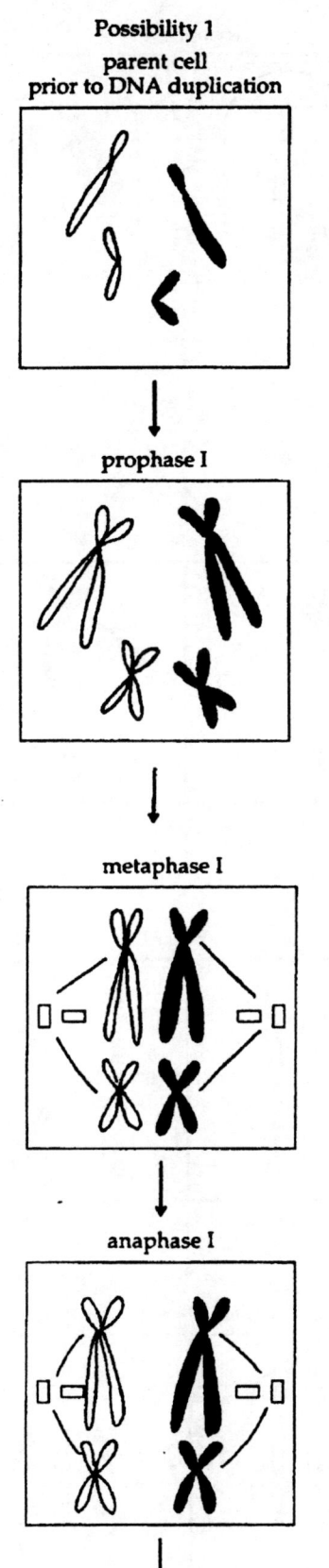

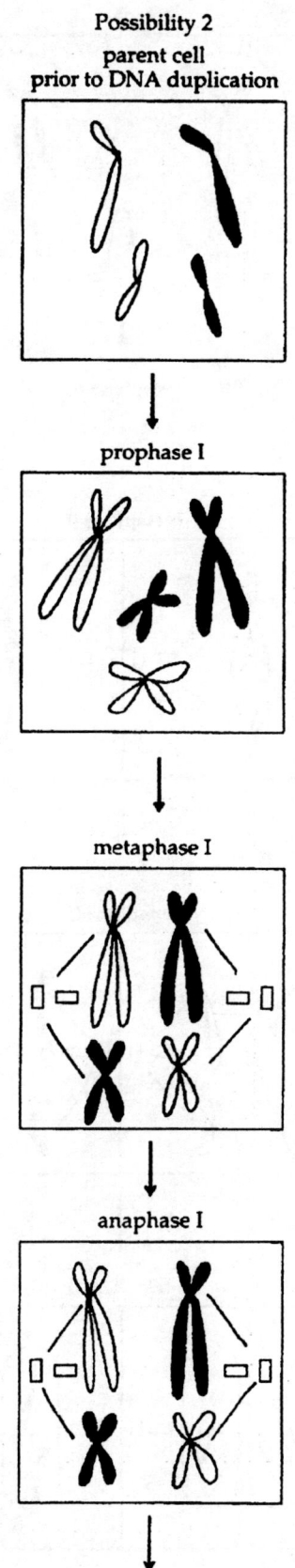

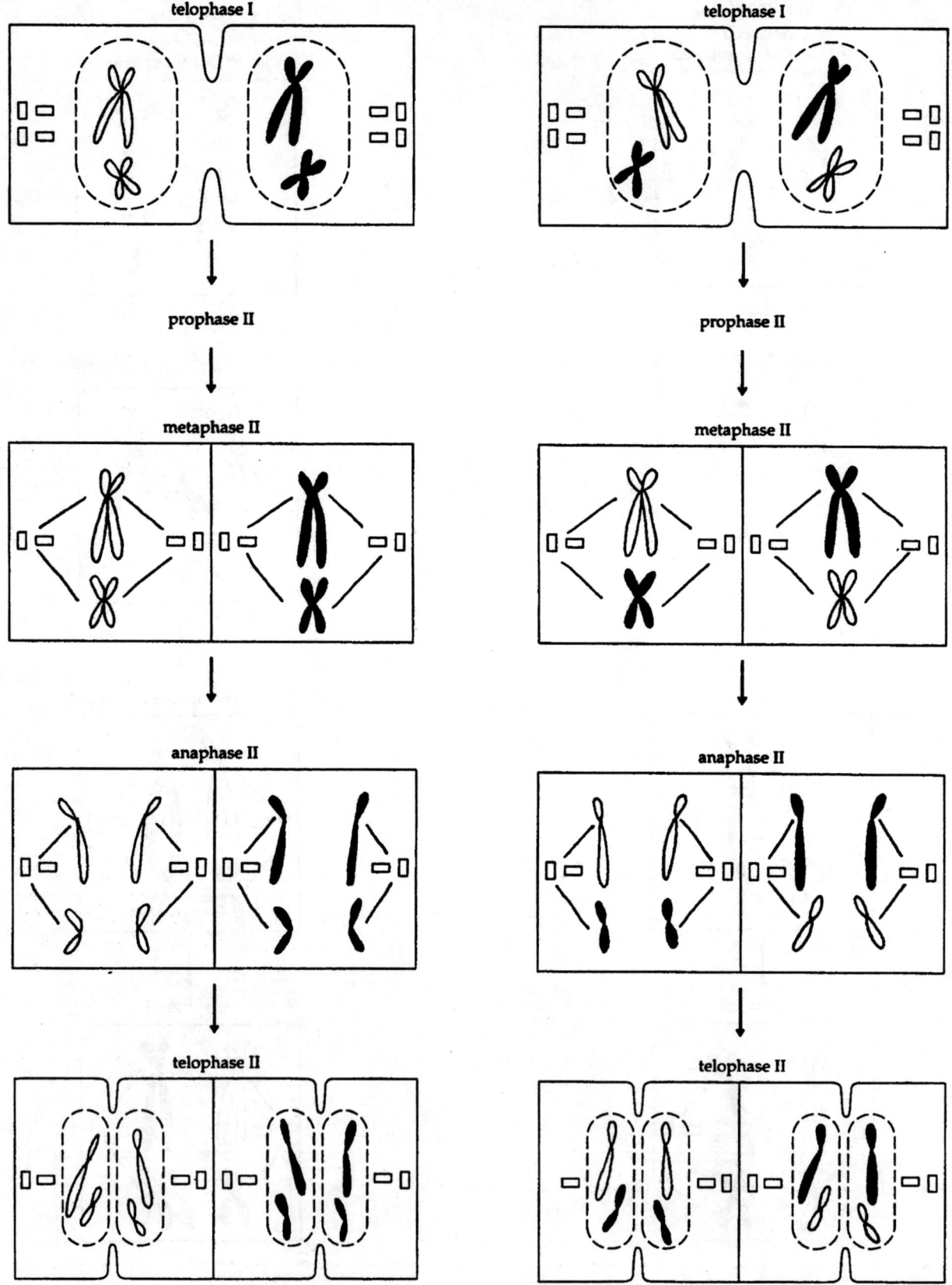

p. 13-16 **b.** 4

      c. 4

      d. *rs* and *RS*

	RS	*RS*
rs	*RrSs*	*RrSs*
rs	*RrSs*	*RrSs*

      e. *RrSs*

      f. purple; smooth

      g. Possible F_1 gametes: *RS; Rs; rS; rs.*

p. 13-16 **h.** 9

	RS	*Rs*	*rS*	*rs*
RS	RRSS	RRSs	RrSS	RrSs
Rs	RRSs	RRss	RrSs	Rrss
rS	RrSS	RrSs	rrSS	rrSs
rs	RrSs	Rrss	rrSs	rrss

	purple smooth
	yellow smooth
	purple wrinkled
	yellow wrinkled

p. 13-17 **i.** 9 purple smooth; 3 purple wrinkled; 3 yellow smooth; 1 yellow wrinkled.

       Corn F_2 dihybrid cross (sample student data):

          Observed: 136 purple smooth; 40 purple wrinkled; 36 yellow smooth; 8 yellow wrinkled.

          Total = 220.

          Expected: 124 purple smooth; 41 purple wrinkled; 41 yellow smooth; 14 yellow wrinkled.

      j. Yes.

      k. No.

      l. No. Genes would not assort independently

Extending Your Investigation: Meiosis and Linked Genes

p. 13-17   HYPOTHESIS: If genes *A* and *B* are linked and no crossing-over occurs, then two *AaBb* individuals will produce *AABB*, *aabb*, or *AaBb* in a 1:2:1 genotypic ratio and 3:1 phenotypic ratio.

           NULL HYPOTHESIS: The fact that genes *A* and *B* are linked on the same chromosome will make no difference in the ratio of offspring.

           Prediction—offspring will be in a 1:2:1 ratio and 3:1 phenotypic ratio instead of 9:3:3:1 typical of crosses between heterozygotes for which genes *A* and *B* are unlinked

           Independent variable—genotypes of parents.

           Dependent variable—number of offspring.

p. 13-18   PROCEDURE:

          1. Take two yellow strands (2 chromatids of one yellow chromosome) and two red strands (two chromatids of one red chromosome) of beads representing maternal and paternal contributions to the *AaBb* hybrid. Place letters AB on

each yellow strand and letters ab on each red strand (you cannot have Ab or aB on strands because the original parents were *AABB* and *aabb*.

2. Carry out the process of meiosis to form gametes.
3. Design a Punnet square to show how gametes would combine to form offspring.
4. Now take beads and form *AaBb* individuals in which *A* and *B* are on separate chromosomes. Use long and short yellow strands as maternal and long and short red strands as paternal. Place letters A on one yellow short strand and a on the red short strand (or vice versa). Place B on the yellow long strand and b on the yellow long strand. Carry out meiosis to make gametes.
5. Design a Punnet square to show how gametes would combine to form offspring.
6. Compare ratios of genotypes of offspring in the two Punnet squares.

RESULTS:

	AB	ab	AB	ab
AB	AABB	AaBb	AaBb	AaBb
ab	AaBb	aabb	AaBb	abab
AB	AABB	AaBb	AABB	AaBb
ab	AaBb	aabb	AaBb	aabb

1:2:1 genotype
3:1 phenotype

	Ab	aB	AB	ab
AB	AABb	AaBB	AABB	AaBb
Ab	AAbb	AaBb	AABb	Aabb
aB	AaBb	aaBB	AaBB	aaBB
ab	Aabb	aaBb	AaBb	aabb

4:2:2:2:2:1:1:1:1 genotype
9:3:3:1 phenotype

The genotypes in the F_2 generation are 1:2:1.
Results support the hypothesis.
Results allow the null hypothesis to be rejected.
The prediction was correct.
Conclusion—linked genes do not assort independently and, therefore, produce phenotypic and genotypic ratios different from what would be expected if genes are unlinked and assort independently.

EXERCISE D   Meiosis and Crossing-Over in *Sordaria (Optional)*

p. 13-23
a. $g^+ t^+$ (these are called black or simply "+" spores).
b. $g^+ t^-$ (these are called tan or simply "*t*" spores).
c. Haploid mycelia, $g^+ t^+$ and $g^+ t^-$, fuse together to produce a diploid nucleus, $g^+ t^+ / g^+ t^-$.
DNA duplicates in interphase of meiosis I.
Final products of meiosis II are four 1n cells: $g^+ t^+ / g^+ t^+ / g^+ t^- / g^+ t^-$ or ++tt. These haploid cells divide by mitosis to produce eight 1n cells, linearly arranged in the ascus:

+ + + + *t t t t*

d. Non-hybrid with all one color of ascospores because hyphae of one strain mated with hyphae of the same strain.
e. During development, one spore slips by the other.

f. The difference between MI and MII asci is that MI asci result from hybridization (black and tan strain hyphae fuse) but no crossing-over takes place among chromosomes. You can predict the final ascospore arrangement by knowing how chromosomes are aligned in MII asci. In MII asci, hybridization has occurred, but crossing-over also occurs among chromosomes so you could not predict the final arrangement of ascospores until you know how chromatids become aligned and separated in meiosis II.

p. 13-24   Figure 13D-7

Laboratory Review Questions and Problems

1. Prophase I; Metaphase II; Interphase I; Telophase I; Anaphase I; Anaphase II; Prophase I.

2. a. Homologous chromosomes synapse to form tetrads in meiosis I and not in mitosis.
    b. Homologous pairs line up in the center of the cell in Metaphase I of meiosis, but chromosomes line up in single file in Metaphase of Mitosis.
    c. Homologous chromosomes are pulled to opposite ends of the cell in Anaphase I of meiosis, resulting in daughter cells with different genetic make-ups when compared with the parent cells. In mitosis, the chromatics of homologues are pulled apart during Anaphase, resulting in daughter cells with the same genetic makeup as the parents.
    d. DNA does not replicate in Interphase II of meiosis.
    e. The end product of meiosis is 4, l*n* cells. The end product of mitosis is two, 2*n* cells.

3.

		Number of chromatids per chromosome	Number of chromosomes per cell
G_1	Interphase	1	2
S	preceding	2	2
G_2	meiosis	2	2
Prophase I		2 (tetrads)	2
Metaphase I		2	2
Anaphase I		2	2
Telophase I		2	1
Interphase II		2	1
Prophase II		2	1
Metaphase II		2	2
Anaphase II		1	2
Telophase II		1	1

4. If alleles did not segregate (assuming that the alleles we are considering are for the same trait), this would mean that homologous chromosomes had not separated and the zygote would have an extra dosage of genetic material. If no segregation of alleles were possible, gametes would always be the same as the parent cells and no diversity in the genetic makeup of gametes would be realized.

---

5. There are 2^{23} different ways in which 23 pairs of chromosomes can be arranged at metaphase of meiosis.

6. Crossing-over provides for a greater range of genetic diversity by forming different combinations of maternal and paternal genes (recombination).

7. The new combinations of genes provided by genetic recombination may impart advantages to the offspring that help them adapt and reproduce successfully. This ensures that the genes of the individuals are included in the gene pool for the next generation, providing the material for evolution.

8. Two *RR* individuals and two *Rr* individuals are produced. All plants are purple. Segregation of alleles occurred. Independent assortment of chromosomes occurred.

	R	R
R	RR	RR
r	Rr	rr

9. Two pea plants, *RrSs* and *rrSs,* are crossed. Three plants are purple smooth, three are yellow smooth, one is purple wrinkled, and one is yellow wrinkled.

	RS	Rs	rS	rs
rS	RrSS	RrSs	rrSS	rrSs
rs	RrSs	Rrss	rrSs	rrss

10. Two genes are linked when they are on the same chromosome and they assort together; they do NOT assort independently. Some genes are linked closely enough together than crossing over does not usually separate them. The differences in frequency with which alleles can be separated by crossing-over can be used to map the genes on chromosomes. (See laboratory 14.)

11. Crossing-over between chromatids of the two chromosomes could produce new genetic combinations that lead to blue, smooth, short plants. They are, however, rare.

12. a. I and II.
    b. I and II.
    c. III, IV, V, VI
    d. Because the chromosomes line up in different ways and any of the chromatids can cross with each other.
    e. Only two chromatids crossed-over.

13.

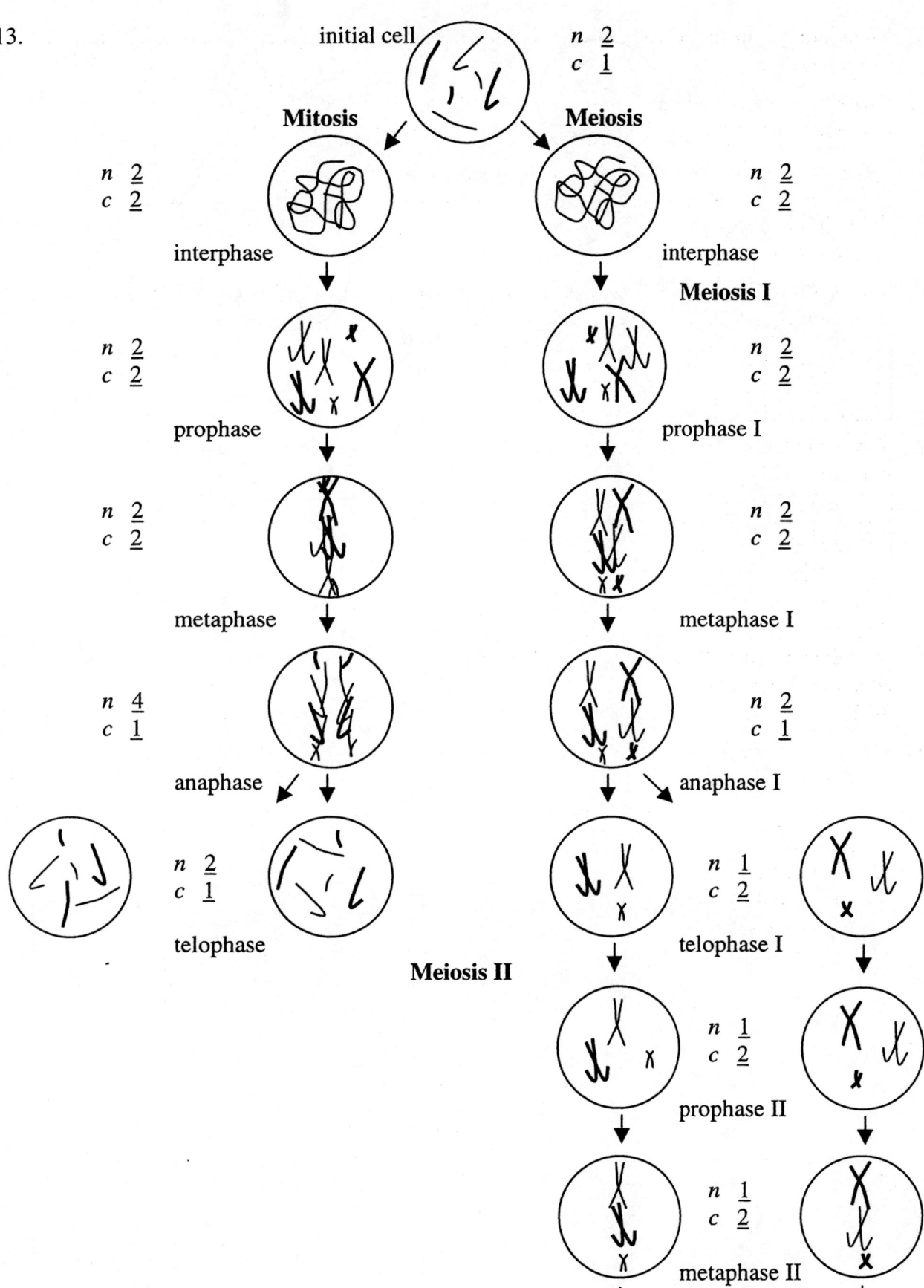

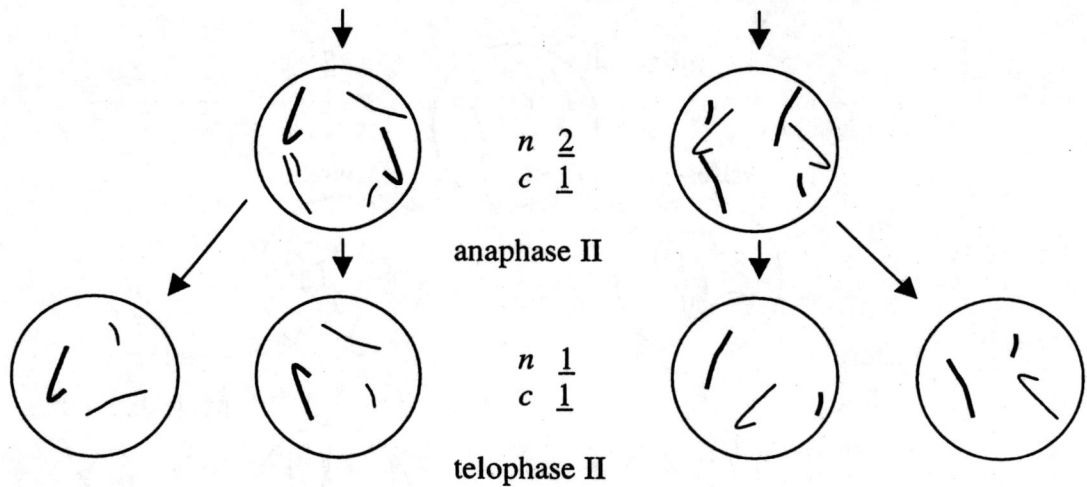

$$n \quad \underline{2}$$
$$c \quad \underline{1}$$

anaphase II

$$n \quad \underline{1}$$
$$c \quad \underline{1}$$

telophase II

Laboratory 14

Genes and Chromosomes: Chromosome Mapping

PART 1   Mapping *Sordaria* Chromosomes (Week 2; see Laboratory 13, Exercise D)

p. 14-2   1. **Table 14A-1 Class Data**

Total asci counted	290
Total MI asci	143
Total MII asci	147

a. 15.3 map units [(147/290) = 50.68%/2 = 25.3 map units].

p. 14-3   6. **Table 14A-2 Your Data**

Number of MI Asci Showing No Crossover	Number of MII Asci Showing Crossover	Total MI + MII Asci	Percentage of Asci Showing Crossover	Frequency/2 (map units)
12	3			
17	2	50	42%	21
29	10			
	6			

p. 14-4   b. Only 2 of the 4 chromatids of the two chromosomes were involved in crossing-over in one crossing-over event. Only 2 of the 4 products of meiosis, therefore, show recombination in one crossing-over event. Each ascospore results from a mitosis following meiosis. Therefore, only 4 of the 8 resulting ascospores, or ½ of the offspring show recombination. Thus, the frequency of recombination is ½ the frequency of asci showing the MII pattern.

9. 21 map units.

c. The result of 21 map units is not very close to the expected of 26. This is because the sample is small.

PART 2   The Chi-Square Test (Optional)

p. 14-4   1. **Table 14A-3 Class Data**

Number of MI Asci	Number of MII Asci	Total MI + MII Asci	Percentage of Asci Showing Crossover	Frequency/2 (map units)
332	348	680	51.8%	25.6

p. 14-5   **a.** χ^2 **for class simulation**

$\chi^2 = \Sigma$ (Observed – Hypothetical)^2 / Hypothetical

    where:

        Observed = the observed number of crossover asci

        Hypothetical = the expected number of crossover asci.

Since the hypothetical map distance = 26 map units, and this results from dividing percent crossovers by 2, then 26 × 2 = 52% and 52% of the asci should show crossover.

Expected = 52% crossover; 48% non-crossover

    0.48 × 290 = 139.2 = 139 expected non-crossovers (MI)

    0.52 × 290 = 150.8 = 151 expected crossovers (MII)

χ^2 = [(observed non-crossovers – 139)^2/139] + [(observed crossovers – 151)^2/151]

    χ^2 = [(143 – 139)^2/139] + [(147 – 151)^2/151]

    $\chi 2$ = 16/139 + 16/151

    χ^2 = 0.115 + 0.106 = 0.221, df = 1, p = 0.05.

Since the χ^2 value of 0.221 is smaller than the critical value of 3.84, we accept the null hypothesis that there is no difference (or that any difference might be due to chance alone) between the expected map distance of 26 and the calculated distance of 25.3 map units.

  3. χ^2 **for class data**

$\chi^2 = \Sigma$ (Observed – Hypothetical)^2 / Hypothetical

    where:

        Observed = the observed number of crossover asci

        Hypothetical = the expected number of crossover asci.

Since the hypothetical map distance = 26 map units, then 52% of the asci should show crossover.

Expected = 52% crossover; 48% non-crossover

Total number of asci counted = 680

    0.48 × 680 = 326 asci showing non-crossover (MI)

    0.52 × 680 = 354 asci expected crossover (MII)

χ^2 = [(observed non-crossovers – 326)^2/326] + [(observed crossovers – 354)^2/354]

χ^2 = [(332 – 326)^2/326] + [(348 – 354)^2/354]

χ^2 = (6)^2/326 + (–6)^2/354

χ^2 = 0.110 + 0.102 = 0.212

With 1 degree of freedom, there is almost 90% (9 in 10) probability that the differences are due to chance alone, so the null hypothesis is accepted; there is no significant difference between the calculated map distance 25.6 and the expected map distance of 26.

p. 14-6   **b.** Null hypothesis: There is no difference between the predicted map distance of 26 units and the map distance calculated in class.

    Alternative hypothesis: Map distance is not 26 map units.

    c. The class data of 15.6 map units is not significantly different (according to χ^2 statistics) from the expected map distance of 26 map units.

    d. Accept the null hypothesis.

    e. The tan gene is approximately 26 map units from the centromere in *Sordaria*.

---

p. 14-8 **a.** Yes. There are chromosomal "puffs" which represent places where the DNA (which is normally tightly coiled into bands on the chromosomes) unfolds and allows transcription of messenger RNA (mRNA) to take place.

        b. There appear to be 4 chromosomes, but there are actually 4 pairs of chromosomes since the homologs are synapsed along their entire length.

EXERCISE C   **Mapping the Chromosomes of *Escherichia coli***

p. 14-10 **a.** Grow them on plates containing streptomycin (the Str^S pro$^+$ leu$^+$ thi$^+$ thr$^+$ strain would die) or on plates lacking one of the essential vitamins or amino acids, pro, leu, thi, tier, (the Str^S pro$^-$ leu$^-$ thi$^-$ thr$^-$ strain would die).

        b. Look for cells which are resistant to streptomycin and can survive without some externally supplied amino acids.

        c. Thiamine is 80 minutes from the origin so there is little chance that it will be transferred from the Hfr bacteria to the recipients. Since the recipients cannot make thiamine, it needs to be in all of the plates.

p. 14-11 **d.** No. The donor strain is sensitive to streptomycin.

        e. Yes. The plate lacks streptomycin and the donor can make all of the essential nutrients.

        f. Comparative controls, to make sure that the donor strain remains streptomycin sensitive so that recombinant cells can be selected on minimal medium containing streptomycin; donor cells cannot grow on this medium.

        g. No. The recipient strain requires external supplementation with amino acids which are absent in the STR/THI plate.

        h. Yes. The plate contains the required amino acids and vitamin.

        i. Comparative controls, to make sure that the mutant recipient cells remain unable to synthesize pro, leu, thi, and thr and have not spontaneously reverted to wild type for any of these genes.

        j. Yes. The D × R strain grows on the STR/THI medium, indicating an ability to synthesize the amino acids lacking in the STR/THI medium.

p. 14-12 **Table 14C-1 Record of Growth (√) and No Growth (0) of *E. coli***

	Donor	**Recipient**	**D × R 1:10**	**D × R 1:10**
STR/THI	√	0	√	√
M/PLTT	√	√		

        k. pro$^+$ leu$^+$ thr$^+$.

        l. The gene for thiamine is the last to be transferred (after 80 minutes). Since conjugation in class occurred for only 30 minutes, the D × R strain cannot receive the thi$^+$ gene.

        m. Recombinant cells will grow on the medium which contains streptomycin and lacks the amino acids, proline, threonine, and leucine, but donor cells will not grow on medium containing streptomycin; only recombinant cells will grow and can therefore be "selected."

        n. No.

        o. Yes.

p. Yes.

q. The genes for proline, leucine, and threonine are located near the leading point of the chromosome. The gene for thiamine is located near the trailing end of the chromosome and the gene for methionine is located midway.

EXERCISE D   Restriction Endonucleases: Mapping Bacteriophage Lambda

p. 14-16   Table 14D-1

*Hind*III Fragment (bp)	Distance Traveled (cm)
23,130	0.5
9,416	1.0
6,557	1.7
4,361	2.4
2,322	3.3
2,027	4.0

Figure 14D-4a *Number of base pairs vs. distance migrated (see the next page for a detailed plot including* Hind*III and* EcoRI*)*

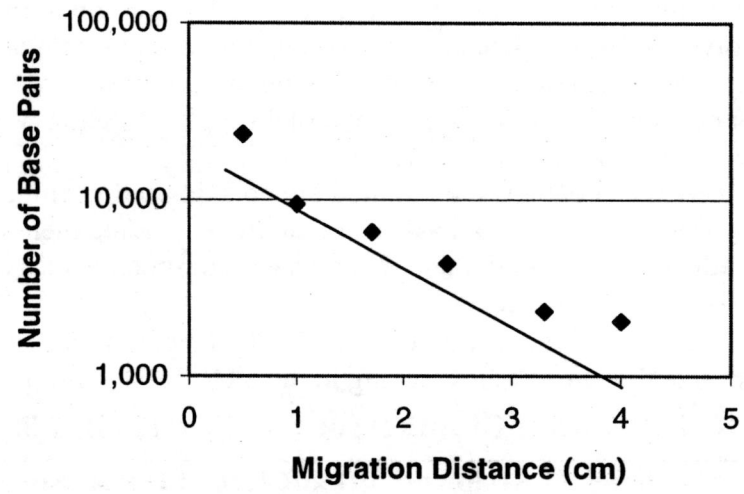

p. 14-19   **a.** Smaller fragments travel farthest through the agarose gel.

Table 14D-2 *Eco*RI Fragment Sizes for Phage Lambda DNA

Expected	Observed	
21,226	not observed—probably within the diffuse band of uncut DNA	0 cm
7,421	7,600	1.5 cm
5,643	5,100	2.2 cm
4,873	4,100	2.6 cm
3,540	3,100	3.1 cm

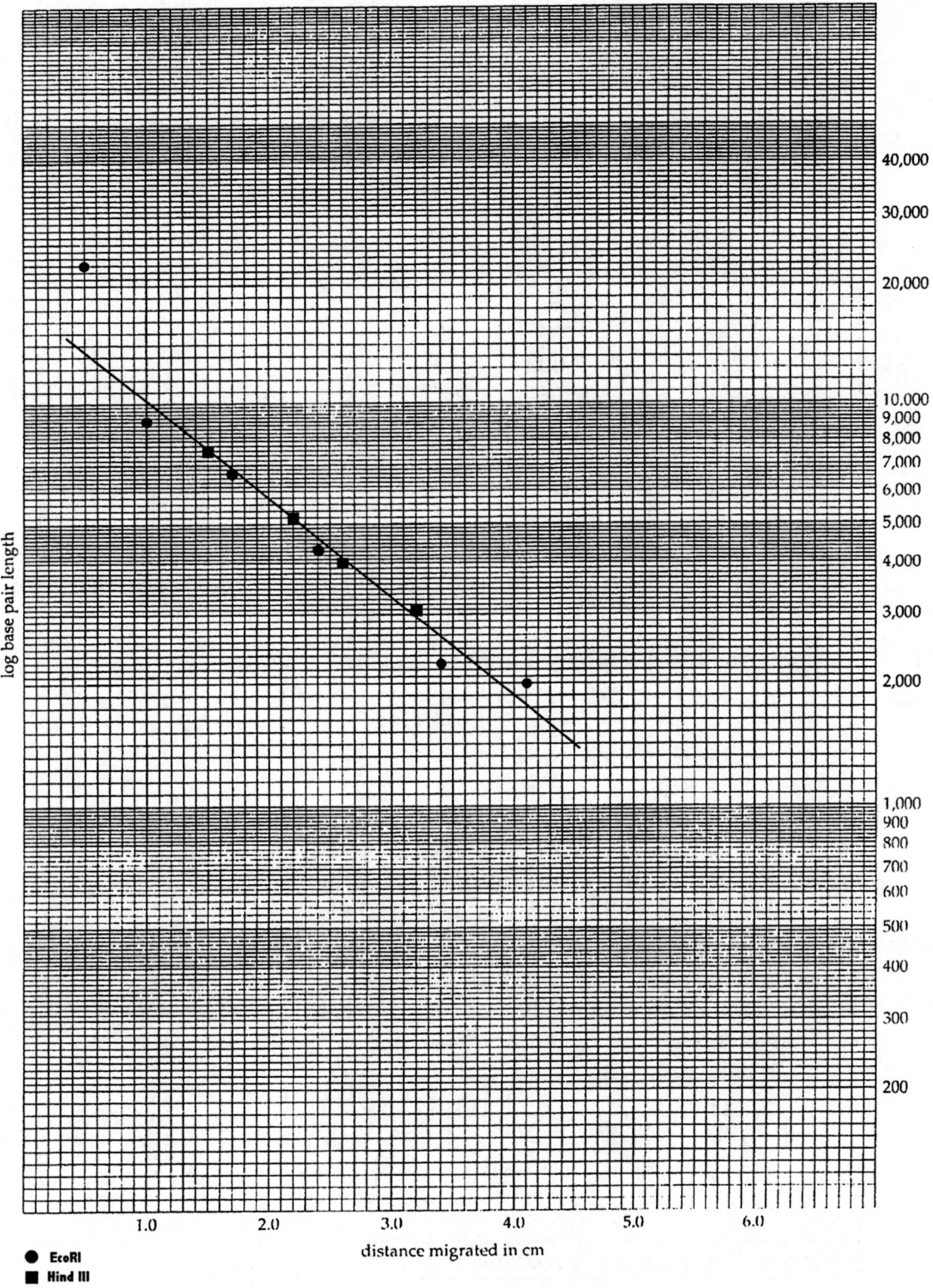

Figure 14D-4b *Semilog graph paper.*

PART 2 Constructing Restriction Maps

p. 14-20   **a.**   2.

      2.

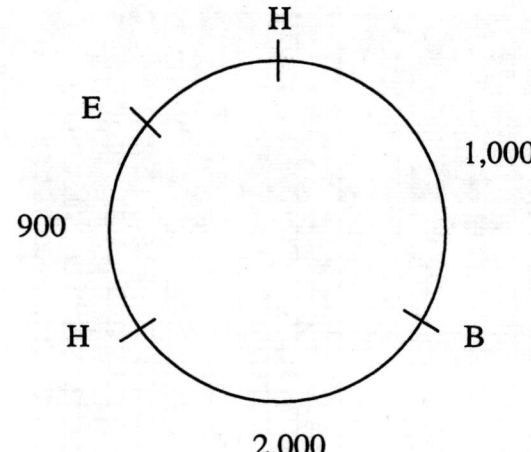

p. 14-21   **b.**   1.

5 and 6.

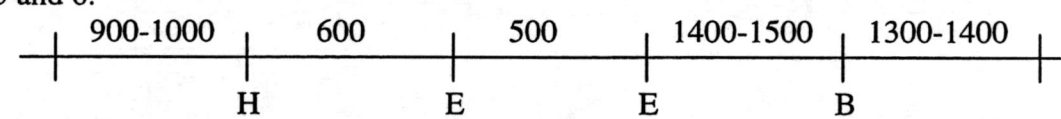

Numbers are approximate. Some bands may move slower because they are in outer wells or may be more diffuse and contain more material.

PART 3 Mapping the Bacteriophage Lambda (λ) Chromosome

p. 14-22   1.

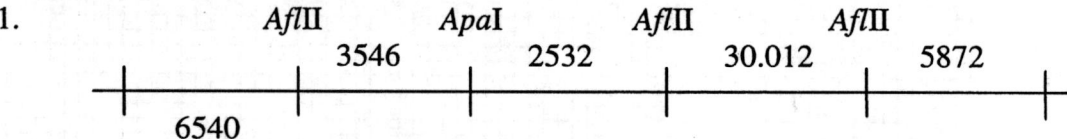

2. *Hind*III sites (kb), base pairs (reading left to right)—
    23,130; 2,027; 2,322; 9,416; 564; 125; 6,557; 4,361
*Eco*RI sites (kb), base pairs (reading left to right)—
    21,226; 4,878; 5,643; 7,421; 5,804; 3,530

p. 14-23   **a.** It appears that bands of 5643 and 5804 base pairs must be together on the gel.
     -   since their size is so similar, their bands appear as one band.

     b. The sizes are approximately the same. The bands for larger fragments appear to be overestimated while those representing small bands appear to be underestimated.

     4.

			↓ travel together ↓			
21226	4878	5643	7421	5804	3530	Actual
21226	4100	5100	7600	5100	3100	Experimental

Laboratory Review Questions and Problems

1. Since, in a haploid organism, only one allele is present at each locus, the phenotypes of individual spores are identical to their genotypes. All alleles are expressed. None are masked

---

by the presence of a different allele at the same locus on a homologous chromosome.

2. The map distance is 11 map units.

Calculations:

Frequency of crossovers = [225 / (800+225)] × 100% = 22%

Map units = frequency of crossovers = 22 map units.

3. A – D = 2 units; D – B = 4 units; B – C = 10 units.

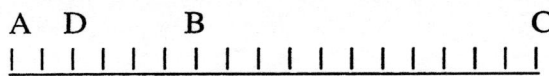

A D   B   C

4. B – C = 6 units; C – A = 4 units; A – D = 9.5 units.

B   C   A   D

total length mapped = 19 units

5.

Genes transferred	Transfer time (min.)
leu$^+$	5
bio$^+$	10
pro$^+$	30
thi$^+$	45

map   F   thi+   pro+   bio+   leu+   F →

6. Since the phosphates that form the backbone of the DNA molecule are negatively charged at neutral pH, DNA will migrate through the gel towards the positive electrode, and thus should be loaded into wells at the negative electrode.

7. Restriction endonucleases protect bacteria by breaking and destroying invading DNA, such as that of bacteriophage viruses. Bacteria protect themselves from their own restriction endonucleases by the production of enzymes that methylate bacterial DNA at each of the specific endonuclease recognition sites. Methylation of a nucleic acid base prevents a close association from forming between the restriction endonuclease and the recognition site, and thus "protects" the bacterial DNA.

8.

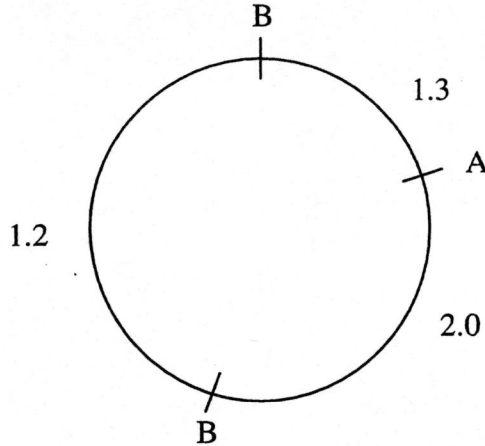

9. Fragment sizes and migration distances:

*Hind*III	cm
23,130	1.2
9,416	1.8
6,557	2.15
3,000	3.28
2,322	3.6
2,027	3.92
725	5.35

Fragments (migration distances)

A + B	cm	A	cm	B	cm
4,500	2.7	7,100	2.0	6,200	2.2
2,800	3.4				
1,600	4.1	1,600	4.1	2,800	3.4

Chromosome fragment arrangement

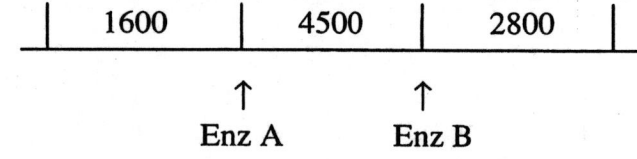

Laboratory 15

Human Genetic Traits

EXERCISE A   Human Cytogenetics

p. 15-2   a. Male

EXERCISE B   X and Y Chromosomes

p. 15-4   a. One

EXERCISE C   Mendelian Inheritance in Humans

PART 1   Monohybrid Crosses in Humans

p. 15-6
    a. Yes, for most traits.
    b. Dimpled chin.
    c. The abundance of the dominant allele (D) for dimpled chin must be low in this laboratory section. Such discrepancies in gene frequency may result from localized concentrations of various ethnic groups, each of which will display a unique pattern of allele frequencies.

p. 15-7   Table 15C-1 Genetic Traits

Characteristic	Your Phenotype	Your Possible Genotypes	Data for Your Laboratory Section	
			Number	Percentage
1. Dimpled (*D*)		*D–*	3/28	10.7
Nondimpled (*d*)		*dd*	25/28	89.3
2. Free ear lobes (*E*)		*E-*	14/28	50
Attached (*e*)		*ee*	14/28	50
3. Widow's peak (*W*)		*W–*	10/28	35.7
No widow's peak (*w*)		*ww*	18/28	64.3
4. Taster of PTC (*T*)		*T–*	18/28	64.3
Non-taster of PTC (*t*)		*tt*	10/28	35.7
5. Left thumb on top (*F*)		*F–*	18/28	64.3
Right thumb on top (*f*)		*ff*	10/28	34.7
6. Bent little finger (*B*)		*B–*	16/28	57
Finger not bent (*b*)		*bb*	12/28	43
7. Hitchhiker's thumb (*h*)		*hh*	17/28	60.7
Normal thumb (*H*)		*H–*	11/28	39.3
8. Long palmer muscle (*l*)		*ll*	13/28	53.6
Two tendons only (*L*)		*L–*	15/28	46.4

Table 15C-1 Genetic Traits (continued)

Characteristic	Your Phenotype	Your Possible Genotypes	Data for Your Laboratory Section	
			Number	Percentage
9. Pigmented iris (P)		pp	15/28	53.6
Non-pigmented iris (p)		P-	13/28	46.4
10. Mid-digital hair (M)		M—	16/28	57.1
No mid-digital hair (m)		mm	12/28	42.9
11. Shorter second finger–female (S^S)		S^S 71.9	male	65.6
Longer second finger–female (S^L)		s^l 25	female	50

PART 2 How Individual Is Each Individual?

p. 15-7 **a.** The number of characteristics necessary to define an "individual" is dependent upon the degree of heterogeneity in the class and the individuals present.

b. Later.

p. 15-8 **c.** Free earlobes, Tastes PTC

d. *ET*

e. Attached earlobes, Non-taster of PTC

f. *et*

g. Genotypes: *EeTt*. Phenotypes: Free earlobes, Tastes PTC

h. Each *EeTt* parent potentially produces the gametes: *ET, eT, Et, et*

Construct a Punnet Square to determine the potential genotypes of the offspring:

	ET	*eT*	*ET*	*et*
ET	*EETT*	*EeTT*	*EETT*	*EeTt*
eT	*EeTT*	*eeTT*	*EeTT*	*eeTt*
Et	*EETt*	*EeTt*	*EETt*	*Eett*
et	*EeTt*	*eeTt*	*EeTt*	*eett*

Offspring Phenotypes: Free earlobes, Taste PTC (9); Free earlobes, Non-tasters of PTC (3); Attached earlobes, Tastes PTC (3); Attached earlobes, Non-tasters of PTC (1).

EXERCISE D Chromosomal Abnormalities—Nondisjunction and Translocation

PART 1 Non disjunction: Sex Chromosomes

p. 15-9 **a.** Female

b. One

c. Male.

d. None.

p. 15-9   **Table 15D-1 Nondisjunction in a Human Female**

(egg)	(sperm)	(genotype)	Expected Sex	Number of Barr Bodies	Name of Syndrome
XX	X	XXX	female	2	Triple-X
XX	Y	XXY	male	1	Kleinfelter's
0	X	OX	female	0	Turner's
0	Y	OY	male	0	non-viable

p. 15-10   **Table 15D-2 Nondisjunction in a Human Male**

(egg)	(sperm)	(genotype)	Expected Sex	Number of Barr Bodies	Name of Syndrome
X	XX	XXX	female	2	Triple-X
X	XY	XXY	male	1	Kleinfelter's
X	0	X0	female	1	Turner's
X	YY	XYY	male	0	Jacob's

PART 2   Nondisjunction: Autosomes

p. 15-10   **a.** Chromosomes pair in prophase of oögenesis but do not separate until the time when the egg is ovulated. This is years later (12 to 50 years). Chromosomes synapsed for this length of time may be less likely to separate from one another. Also, older women are more likely to have been exposed to chemical mutagens or radiation which would have an effect on the mechanism of meiosis.

EXERCISE E   Constructing a Human Pedigree

p. 15-11   **2.** Mrs. S's parents were probably a normal father but a mother carrying fragileX. Mrs. S received the fragile X chromosome from her mother but a normal chromosome from her father so she too was a carrier

p. 15-13   **3.** Mrs. S's Family Pedigree:

Potential Grandparents

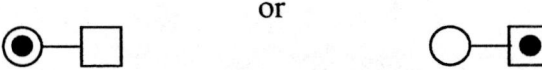

or

Assume that the grandfather did not carry the fragile X chromosome, since if he did he would have been severely retarded. Therefore, the grandmother must have been a carrier.

Mrs. S's grandparents:

Mrs. S's parents and
    siblings

       Mrs. S

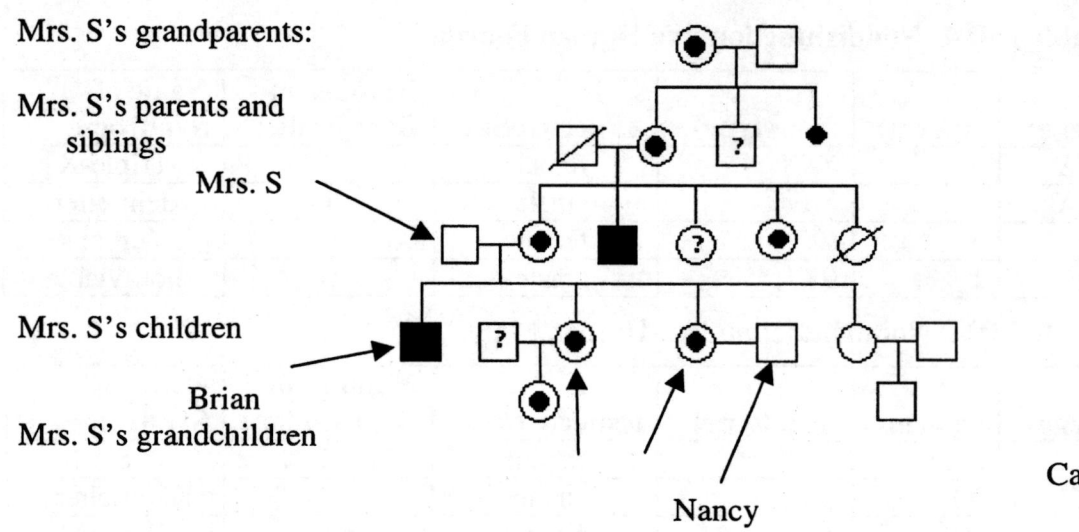

Mrs. S's children

    Brian
Mrs. S's grandchildren

Nancy

Carol   Jane

		Jane	fragile X
		x	X
Jane's	X	Xx	XX
husband	Y	xY	XY

a. Assume that Jane or Carol married normal males. Their chances of producing a son affected by the fragile X syndrome would be 50%. They would have a 50% chance that any daughter would be a carrier of the fragile X gene.

b. Carol's daughter would also be a carrier.

c. Mrs. S's grandmother was a carrier, but her grandfather was a normal male.

d. It is not known whether Mrs. S's uncle had the fragile X or not.

EXERCISE F   Forensic Science: DNA Fingerprinting

p. 15-17   a. Suspect Y

        b. The DNA fingerprint patterns for identical twins are the same.

        Figure *Forensics gel*

1. Man with attached earlobes: *ee*
   Heterozygous woman: *Ee*

	e	*e*
E	*Ee*	*Ee*
e	*ee*	*ee*

   Offspring: 50% with Free earlobes *(Ee)*; 50% with attached earlobes *(ee)*.

2. **a.** Children are either heterozygous *(Pp)* or homozygous for the normal gene *(pp)*.
   b. Individuals with a *Pp* genotype display polydactyly. Individuals with the *pp* genotype are normal.
   c. The possible genotypes of the children are: *PP, Pp.* or *pp.*
   d. Both the *PP* and *Pp* genotype individuals display polydactyly. The *pp* individuals are normal.

3. When the embryo is multicellular (say 16 cells), one of the *X* chromosomes in each of the 16 cells becomes heterochromatic (shuts down). However, it could be *either X,* so it is just chance as to whether the paternal *X* or maternal *X* (which may carry different alleles) becomes a Barr body. Thus, the 16-cell embryo is a "mosaic."

4.
   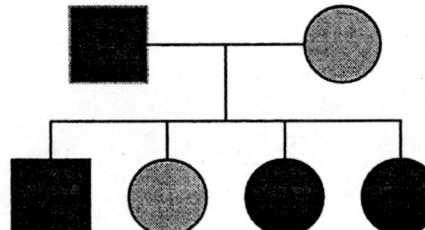

   a) colorblind father   b) heterozygous mother

   c) colorblind son   e), f) colorblind daughters
   d) heterozygous daughter

5. Yes, the mother is heterozygous and contributes the X^c gene to the gamete fertilized by the Y-carrying sperm; thus, a colorblind son results. A colorblind daughter would result from the same type of egg (X^c being fertilized by a X^c-carrying sperm). Probability 0.25 for each case.

6. **a.** I^A, I^B
   b. i, i
   c. $I^A i$ or $I^B i$
   d. Type A ($I^A i$) or type B ($I^B i$).
   e. Type O, Type AB.

7. **a.** No. Parents would not have type I^A alleles.
   b. No, because the two parents may have all of the possible alleles for blood type (I^A, I^B, i)
   c. No, because the father must contribute the type I^A allele and he does not have this allele.
   d. The offspring may be type A ($I^A I^A$), type B ($I^B I^B$) or type AB ($I^B I^A$).

8. No, you cannot rule out any of the three men as possible fathers. Both men with blood types A and B could carry the *i* allele for blood type. In this case, their possible blood types would be $I^A i$ and $I^B i$. The man of blood type O would be *ii*. Either subject 3 or 4 could be the possible father. Hereditary material represented by bands 6 and 8 in the child could not have come from the mother. Both subjects 3 and 4 have these bands. The last band in the sample

from subject 4 does not have to appear in the child.

9. 1. DNA is extracted and treated with restriction enzymes.
   2. Fragments are subjected to electrophoresis.
   3. Denature DNA on gel (heat).
   4. Transfer DNA fragments to nitrocellulose filter paper (blotting).
   5. Heat with a radioactive probe that will bind to certain versions of RFLP markers and not to others.
   6. After the radioactive probe is rinsed off, autoradiography is performed to show the bands that base-paired with the probe. The probability that two people would have the same set of RFLP markers is very small—each person is unique in terms of variations in his or her DNA that changes the length of the restriction fragments, forming different band patterns.

10. DNA could be extracted from the bones. Since this would be found only in small amounts, it would have to be amplified using PCR (polymerase chain reaction). The DNA could then be subjected to fingerprinting (see question 8) and compared to the DNA fingerprints of the closest relatives.

Laboratory 16

DNA Isolation

EXERCISE A   DNA Isolation Procedure

PART 1   DNA Isolation (Chloroform)

p. 16-2   a. In the filtered homogenate.

p. 16-3   b. Denatured proteins.

p. 16-4   c. Rotating the rod in the same direction will assure the continuous winding of the long DNA strands onto the rod.

PART 2   DNA Isolation (No Chloroform)

p. 16-4   a. Yellowish.

           b. Meat tenderizer denatures the proteins complexed with DNA.

p. 16-5   c. The DNA is a very long polymer of nucleic acids and is usually stringy and thread-like if isolated without damage.

EXERCISE B   Isolation of DNA from Animal Cells (*Optional*)

PART 1   Isolation of DNA

p. 16-5   a. Yes. The viscosity increased and the color of the contents lightened as proteins were denatured and precipitated.

PART 2   Colorimetric Detection of DNA

p. 16-6   7. Color: tube I = blue (though a different color blue than the standard, since this isolated material also contains associated protein and RNA); DNA standard = blue;
Control (water) = clear.

           a. To demonstrate the color of diphenylamine in the presence of pure DNA.

           b. Since the isolated material is not pure DNA, but rather contains associated proteins and RNA, the reaction with diphenylamine reagent will produce a blue color that is an inexact match to the color of diphenylamine in the presence of pure DNA. Therefore, both the DNA standard and the water control are compared to the isolated material, to demonstrate the presence of DNA.

EXERCISE C   Isolation of DNA from Bacteria (*Optional*)

PART 1   Isolation of DNA

PART 2   Measuring Absorbance of DNA

p. 16-7   a. The isolated DNA contains other UV-absorbing molecules.

           b. The A280/260 ratios are similar for the two preparations. The curve for the DNA standard with a peak at 260 is typical for pure preparations. The preparation of DNA isolated from E. *coli* contains low concentrations of A_{280} - absorbing proteins (though it does contain proteins and other contaminants that absorb at lower wavelengths.

PART 3   Colorimetric Detection of DNA

PART 4   Preparing a Standard Curve

p. 16-8   DNA concentration = 0.0049 mg/ml
(Total DNA extracted from 2 g (wet weight) of bacterial paste = 0.392 mg)

Laboratory Review Questions and Problems

1. DNA is a long chain polymer which is ubiquitous in all living organisms (as demonstrated by the extractions from prokaryotes, plants and animals).

2. The genetic material must contain the codes for all molecules produced in a living organism, and thus must allow for a great number of variations in molecular composition. When it was discovered that DNA contained only four nucleotides, many questioned how such lack of variation could account for the variations found in living organisms. It was thought that since proteins are so diverse, they might be better candidates to serve as genetic material. However, we now know that the four nucleotides are used three at a time (a codon) to code for sequences of amino acids. This allows for 4^3 or 64 different combinations of three nucleotides (or 64 possible codons). These can easily be used to code for 20 amino acids. Indeed, most amino acids are coded for by more than one codon (the genetic code is "degenerate"). Depending on the order of codons, the order of amino acids in a polypeptide chain will vary. Various organisms contain 5,000 (simplest virus) to over 5 billion (human) paired nucleotide bases. Therefore the number of potential base sequences is great.

3. Protein molecules are considerably shorter in length than are DNA molecules. Therefore, it is easier to spool out DNA, even if it has been cut during processing (the pieces are still relatively long). A protein may contain as many as several hundred amino acids. In contrast, DNA strands contain thousands to millions of nucleotide base pairs. The DNA from a single human cell, if measured as a continuous thread, extends almost two meters in length (though in the cell, this length of DNA exists as shorter structures; the individual chromosomes).

4. <u>Step 1:</u>   Heat diced onion tissue in homogenizing medium for 15 min. at 60°C
       Purpose: 1. Softens onion tissue to allow penetration of homogenizing medium;
                       2. Denatures many enzymes;
                       3. Dissolves cell membranes.
     <u>Step 2:</u>   Cool preparation to 15° - 20° C in an ice bath.
       Purpose: Rapid cooling prevents DNA denaturation.
     <u>Step 3:</u>   Homogenize preparation in a blender (45 sec. at low speed; 30 sec. at high speed).
       Purpose: 1. Ruptures cells and releases contents;
                       2. $NaCl_2$/sodium citrate buffer stabilizes the DNA by forming a Na^+ shell around the negatively charged phosphates of the DNA;
                       3. Citrate inactivates DNase.
     <u>Step 3:</u>   Cool homogenate in an ice bath for 15 - 20 min.
       Purpose: Precipitated proteins and empty cell walls separate.
     <u>Step 4:</u>   Filter homogenate.

Purpose: Removal of (some) precipitated proteins and empty cell walls.

Step 4: Pour 50 ml of filtered homogenate into a 250 ml flask and gently add 2 ml chloroform by pouring it down the side of the flask. Gently swirl flask.

Purpose: Precipitation of proteins.

Step 5: Decant only the homogenate (top) layer into another 250 ml flask. Repeat chloroform extraction (Steps 4 and 5) four more times.

Purpose: Precipitation of proteins.

Step 6: After the fifth deproteinization, pour homogenate into a 250 ml beaker. Cool in ice bath until it reaches 10° -15° C.

Purpose: Cooling the DNA suspension aids in the final extraction.

Step 7: Add ice-cold ethanol down the side of the beaker until a DNA precipitate appears.

Purpose: DNA precipitation.

Step 8: Spool DNA onto a scored glass rod by rotating the rod *in one direction only* in the DNA beaker.

Purpose: Collection of precipitated DNA.

5. To determine DNA concentration in the tissue, record the weight of the tissue to be extracted. Complete the appropriate DNA extraction procedure, taking care not to spill or otherwise lose material in the process. Spool out all of the ethanol precipitated DNA. After spooling, the amount of DNA collected must be quantified. Desiccate (dry out) the isolated DNA and weigh it. Redissolve a measured aliquot (weighed amount) of the desiccated DNA in buffer or 4% N NaCl and test with diphenylamine reagent. Compare the color (or absorbance, if a spectrophotometer is available) produced to that of a series of standards, made up of known concentrations of pure DNA reacted with diphenylamine reagent. Graph the absorbance against the DNA concentrations of the standard solutions. From this graph, locate the absorbance of the isolated DNA sample and the corresponding DNA concentration. This is the concentration of DNA in the tested solution, but represents only a portion of the total DNA in the tissue extracted. Calculate the total quantity of DNA in the tissue extracted by comparison with the total amount (weight) of precipitated DNA collected. See Laboratory 4, Exercise C on the use of a standard curve to determine concentration.

Notes:

Laboratory 17

DNA—The Genetic Material: Replication, Transcription, and Translation

EXERCISE A   Replication

p. 17-5   **a.** Cytosine, Thymine, Adenine, Guanine
b. 3'–5' phosphodiester bonds
c. Hydrogen bonds

p. 17-6   **d.** Blue strand I' is identical to white strand II. Blue strand I' is complementary to white strand I.

EXERCISE B   Transcription and Translation

PART 1   Transcription—RNA Synthesis

p. 17-8   **a.** 5' refers to the phosphate group on the #5 carbon of the terminal ribose.
3' refers to the hydroxyl (OH) group on the #3 carbon of the terminal ribose on the opposite end of the RNA strand.
3. 5' end AUGCCGUAUACCUAA 3' end
b. Each codon codes for a specific amino acid and pairs with a complementary anticodon on a tRNA carrying an amino acid.

PART 2   Translation—Protein Synthesis

p. 17-15   18. The completed peptide chain is: Methionine - Proline - Tyrosine - Threonine - STOP.

EXERCISE C   Point Mutations in DNA

PART 1   Base Substitutions—Possible Effects

p. 17-15   The DNA strand used to make the mRNA was: 5' end TTAGGTATACGGCAT 3' end.
a. The sequence of nucleotides in the third codon of the mRNA (after the base substitution) is: 5' end CAU 3' end
(modified DNA is: 5' end TTAGGTATGCGGCAT 3' end).
b. This codon specifies histidine.
c. Potential effects of the base substitution are on protein conformation and interactions with other molecules. The original third amino acid in the chain was tyrosine, an uncharged molecule. Since histidine is a basic amino acid (positively charged at pH 7) substitution of this amino acid into the chain adds a strong positive charge.
d. Changing of the ninth nucleotide from an A to a C will form the codon, 5' end GAU 3' end, which codes for the acidic amino acid aspartic acid (negatively charged at pH 7).

    e. Changing of the ninth nucleotide from an A to a T will form the codon, 5' end AAU 3' end, which codes for the polar, uncharged amino acid asparagine.

    f. Protein conformation and interactions with other molecules of the chain containing asparginine, which is polar uncharged like tyrosine, would be most like the original peptide chain, but all of the changes produce defective proteins.

PART 2   Base Substitution Resulting in Sickle-Cell Anemia

p. 17-16
  a. The codons for glutamic acid are GAA or GAG.

  b. The codons for valine are GUU or GUC or GUA or GUG.

  c. Change the second nucleotide from T to A.

PART 3   Frame-Shift Mutations: Base Additions and Deletions

p. 17-16
    mRNA: 5' AUGCCGUAUACUCUAA 3' (DNA: 5' TTAGAGTATACGGCAT 3')
    Amino Acids: methionine - proline - tyrosine - threonine - leucine

p. 17-17
  a. The sequence is the same, with the addition of another amino acid instead of the chain being terminated by the stop codon.

  b. The removal of a nucleotide in the original DNA will change the types of amino acids in the protein produced, and will certainly result in the loss of the last amino acid (or STOP message), since only two amino acids will be available for the codon.

Laboratory Review Questions and Problems

1. A five base code word would code for 32 amino acids.

2. **a.** mRNA: AAU AGA AGC CCU CUC UUU UGU
   protein: Asparagine - Arginine - Serine - Proline - Leucine - Phenylalanine - Cysteine.

   b. After deletion, mRNA: AUA GAA GCC CUC UCU UUU GU?
   protein: Methionine - Glutamic Acid - Alanine - Leucine - Serine - Phenylalanine (next amino acid depends on the next letter)

3.   delete G   insert G
         ↓   ↓

   AAG AGU CCA UCA CUU AAU GCU

   AAA GUC CAU CAC UUA AUG GCU
   |_____↑
         frame shift   corrects itself

4. 21% A; 29% G; 29% C; 21% U

5. **a.** mRNA: 5' AUG CCA UUU UGA 3'
   c. codons
   d. AUG is the initiation codon (fMet or Met).
   e. UGA is the STOP codon

6. **a.** clockwise, from upper right: A site; large ribosomal subunit; small ribosomal subunit; P site.

b. 5' end is on the ribosome (i.e., on the left side of the mRNA.

c. Anticodon

d. 5' CCA 3'

e. peptide bond, peptidyl transferase (although indications are that the ribosome itself may possess proteins that catalyze peptide bond formation).

7. mRNA:

    5' U C U A|A U G|A G C|U C G|G C C|C A U|U A G|C C G 3'

protein:

    fmet – ser – se – ala – his – (stop)

8. Complementary DNA:

    5' T C T A A T G A G C T C G G C C C A T T A G C C G 3'

mRNA made in 5' to 3' direction and read 5" to 3':

    5' C G G C U A|A U G|G G C|C G A|G C U|C A U|U A G|A)

protein:

    f met – gly – arg – ala – his – (stop)

9. fmet – pro – asp – gly – thr

First, arrange all anticodons 3'–5' because they must be antiparallel to codons they match and the codons are 5'–3' because mRNA is read in a 5' to 3' direction. Next, determine 5'3' codons that are complementary to the anticodons.

3' C C G 5'	G G C (gly)	
U G U	A C A (thr)	fmet – pro – asp – gly – thr
G G C	C C G (pro)	AUG CCG GAC GGC ACA
U A C	A U G (fmet)	
C U G	G A C (asp)	
ANTICODONS	CODONS	
3'–5'	5'–3'	

Use the genetic code to determine which codon codes for an amino acid in the original protein and assign that codon to the amino acid.

mRNA:

    5' A U G C C G G A C G G C A C A 3'

DNA:

    3' T A C G G C C T G C C G T G T 5'

    5' A T G C C G G A C G G C A C A 3'

Laboratory 18

Molecular Genetics: Recombinant DNA

EXERCISE A   Bacterial Transformation: Constructing Recombinant Plasmids

p. 18-5
    a. The 3' and 5' ends are composed of complementary base sequences.

    b. *Bam*H1. The plasmid contains a recognition site for *Bam*H1, but does not contain the recognition site for *Eco*R1.

    c. The human DNA sequence to be inserted must be able to be cut with a restriction enzyme that produces sticky ends compatible with those produced by the *Bam*H1 cut in the plasmid.

EXERCISE B   Rapid Colony Transformation with pAMP: Ampicillin Resistance

p. 18-6
    HYPOTHESIS: If the Amp^r gene is transferred to Amp^s bacteria, then the recipient bacteria will be resistant to ampicillin.

    NULL HYPOTHESIS: Presence of the Amp^r gene will not make a difference in ampicillin resistance among Amp^s bacteria.

    Prediction—Transformed bacteria will grow on agar containing ampicillin.

    Independent variable—presence or absence of ampicillin.

    Dependent variable—growth of bacteria.

P. 18-7
    16. LB–   *yes*   LB/Amp–   *no*
        LB+   *yes*   LB/Amp+   yes

    a. Controls. The ability of "–" cells to grow on LB agar (LB– plates) shows that the cells are viable. The inability of "–" cells to grow on LB/Amp agar (LB/Amp-plates) indicates that the cells are not *Ampicillin* resistant and must be transformed (altered genetically) to become ampicillin resistant.

    b. The plasmid which conferred ampicillin resistance was not added to the "–" preparations.

    Results support the hypothesis.

    Results allow the null hypothesis to be rejected.

   - Conclusion—when cells are successful in taking up the pAmp plasmid, they become transformed and are ampicillin resistant

    a. 0.05 μg pAMP

    b. 0.05 μg/510 μl = 0.000098 μg/μl

    c. 0.000098 μg/μl × 100 μl = 0.0098 μg

    d. (student data) 300 – 500 colonies

    e. Transformation efficiency = $3.1 \times 10^4 - 5.1 \times 10^4$ colonies/μg pAMP. The transformation efficiency is determined by the number of competent cells in the preparation and the concentration of plasmid DNA. The calcium chloride procedure for making competent cells generally yields transformation efficiencies of 10^5-10^7. Students typically obtain transformation efficiencies of 10^4-10^5.

p. 18-8   HYPOTHESIS: If the β-galactosidase gene is transferred to lac⁻ bacteria along with the Amp^r gene, then the cells will be resistant to ampicillin and able to catabolize lactose.

p. 18-9   NULL HYPOTHESIS: Presence of the Amp^r and lac⁺ gene will not make a difference in the ampicillin resistance or ability to use lactose among recipient bacteria.

Prediction—transformed cells will grow on agar containing ampicillin and will turn blue if the plates contain X-gal.

Independent variable—presence or absence of ampicillin and X-gal.

Dependent variable—growth and color of bacteria.

a. The (–) pBLU cells serve as a control and must be treated in the same way as the experimentals. If heat shock damages the cells, the fact that the control cells will be unable to grow on LB agar will indicate such is the case.

p. 18-11 b. The LB plates serve as positive controls: growth should be seen for both transformed and non-transformed cells if the cells remain viable. The LB/Amp and LB/Amp X-gal plates inoculated with non-transformed cells serve as negative controls. No growth should occur on these plates because the non-transformed cells are NOT ampicillin resistant.

c. To check for viability.

d. The cells are capable of growing. They have not been harmed by experimental treatments.

e. The cells are not viable. They may have been harmed when heat shocked or from other errors in handling.

Table 18C-1 pBLU Transformation

Plate	Cells	Growth (G) or No Growth (NG)
LB	(+) pBLU	G
LB	(–) pBLU	G
LB/Amp	(+) pBLU	G
LB/Amp	(–) pBLU	NG
LB/Amp/X-gal	(+) pBLU	G
LB/Amp/X-gal	(–) pBLU	NG

†8. 362 colonies.

f. Blue.

g. Yes

p. 18-12 h. Cells appear white because the non-transformed cells lack the pBLU plasmid. They feed on glucose in the medium instead of lactose because they lack a portion of the β-galactosidase gene. Without this gene, the bacteria cannot break down X-gal either (it is X-gal that turns blue). They are able to grow even though ampicillin is present because the cells in the "blue" colony are breaking down the ampicillin in the agar. The "satellite" colonies can grow in these peripheral areas.

i. Yes. Cells did not have to be transformed to grow on LB agar.

j. Cells treated with pBLU plasmid grew on LB/Amp plates, but non-transformed cells were unable to grow on LB/Amp plates. The non-transformed cells did not acquire the plasmid that carries the gene for ampicillin resistance.

  k. The cells grow on the LB/Amp/X-gal plates because they are ampicillin resistant. The pBLU plasmid taken up by the cells carries the Amp^r gene which has been successfully transferred to the ampicillin-sensitive bacteria.
  Results support the hypothesis.
  Results allow the null hypothesis to be rejected.
  Conclusion—cells transformed by uptake of the pBLU plasmid become resistant to ampicillin and are able to use lactose as a food source.

20 a. 0.05 µg pBLU.

  b. 0.05 µg/(250 µl + 10 µl + 150 µl) = 0.000098 µg/µl × 100 µl = 0.0098 µl.

  c. 0.00003 µg/µl × 300 µl = 0.09 µg pUC8

  d. 362 colonies.

  e. 362/0.0098 µg = 3.7 × 10^4 colonies/µg pBLU.

  l. Transformation efficiency is generally lower than for Exercise B.

Laboratory Review Questions and Problems

1. The first consideration would be to find restriction endonucleases specific for nucleotide sequences included within the plasmid and chromosome. The next step would be to cleave the plasmid and the chromosome with restriction endonucleases which provide compatible sticky ends to allow the plasmid to be inserted into the chromosome.

2. Transformation efficiency (number of transformed colonies/µg plasmid DNA)=
   Total number of colonies on the plate = 800 colonies = 800
   Total µg plasmid in reaction mixture 50 µl × (1 × 10^{-3} µg/ml) 5 × 10^{-5} µg
      NOTE: 1 × 10^{-3} µg/ml = 1 × 10^{-6} µg/µl; 50 µl × (1 × 10^{-6} µg/µl) = 5 × 10^{-5} µg
   Transformation efficiency = 1.6 × 10^7 colonies/µg plasmid

3. a. Expect to see both growth on LB(+) and LB(−) plates. No growth on LB/Kan (−). Growth on LB/Kan (+).

   b. Perhaps enzyme A cut in the middle of the Kan^r gene but enzyme B cut at a different place outside the Kan^r gene.

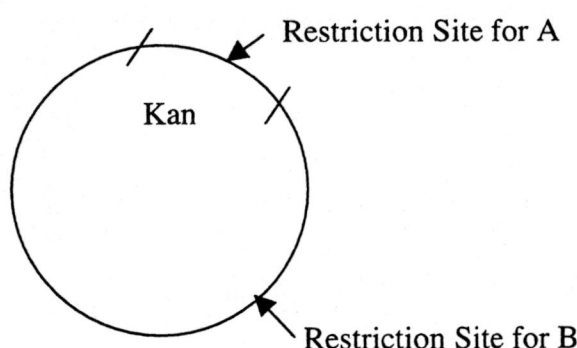

Notes:

Laboratory 19

Genetic Control of Development and Immune Defenses

PART I DEVELOPMENT

EXERCISE A   Fertilization and Early Development in Sea Urchins

EXERCISE B   Cleavage

p. 19-7   **a.** The nucleus contains abundant tRNA and mRNA to allow rapid protein synthesis and cell division without RNA synthesis.

b. No, it is the same.

c. Cleavage diminishes cell size. Embryo size does not change.

p. 19-9   **d.** Yes. The early developmental stages of starfish and sea urchin appear to be the same.

e. Yes, the first cleavage plane cuts the gray crescent of the frog embryo in half, laying down the future right and left sides of the embryo.

5. No, the lower blastomeres are larger. Cleavage of the frog embryo is holoblastic and unequal.

EXERCISE C   Formation of the Blastula

p. 19-10   **a.** The blastula is a hollow sphere, therefore we are looking through two layers of cells in the center and many layers of cells at the periphery.

b. As in (a) above, the apparent cell density is greater at the periphery of the late blastula.

c. The cells of the vegetal pole (yolk) are larger and divide with difficulty. The blastula is solid in the vegetal region. Animal pole cells are smaller and divide easily. The third cleavage division occurs within the animal portion of the embryo. Therefore, the blastocoel is displaced into the animal pole. d. Animal pole blastomeres are smaller.

d. Animal pole blastomeres are smaller.

EXERCISE D   Gastrulation

p. 19-12   **a.** The mesoderm, endoderm and future notochord will differentiate into tissues that occupy the inner body cavity of the mature embryo.

b. The notochord gives rise to intervertebral disks; the notochord disappears in the adult. The mesoderm gives rise to the coelom lining, muscles, organ systems (except the nervous system, which is formed from neural ectoderm) and vertebrae. The endoderm gives rise to the gut, pharynx lining and organ ducts.

c. The paired mesodermal coelomic pouches are continuous with the endoderm of the gut (enteron) during early stages of development. Later they become detached from the gut. Since the mesoderm of the coelom was originally continuous with the endoderm tissue, these organisms are referred to as being "entero-" (gut) "coelous" (body cavity).

EXERCISE E   Neurulation

p. 19-13   a. The embryo elongates.

EXERCISE F   Development of the Chick

EXERCISE G   Formation of Extraembryonic Membranes in the Chick

p. 19-16   6. The chorion appears to extend about half way around the yolk as far as the yolk sac.

PART II IMMUNE RESPONSES

EXERCISE H   Demonstrating the Immune Response by the Precipitin Ring Test

p. 19-19   a. The number of memory cells capable of being stimulated by the antigen is larger, causing the secondary response to set in with greater rapidity, even in response to weak concentrations of antigens. Booster shots are used to bring on this secondary response so that the titer of antibodies in the blood provides adequate protection against a specific infection. Inactivated viruses used as vaccines usually produce an immunity that requires frequent boosters. Attenuated liver viruses used as vaccines produce a more persistent immunity that may last for up to 10 years or more, the typical life span of memory cells.

  b. Memory cells.

  c. Memory B cells are formed in response to antigen. Many of these memory cells migrate from the spleen and lymph nodes to colonize the bone marrow. They bear the same antibodies as those produced by plasma cells in response to antigen. When they encounter antigen during subsequent infections, they immediately secrete antibodies. Thus, the lag time of response is shorter and the actual response is stronger.

p. 19-20   Table 19H-1 Ring Test for Detection of an Antigen (BSA) or an Antibody (anti-BSA)

	Tube number				
	1	2	3	4	5
Top layer	B	Ag	B	Ab	Ab
Bottom layer	Ab	B	N	N	Ab
Reaction	none	none	none	none	ring

  d. Tube 1. Antibodies are present against BSA but no antigens (BSA) have been introduced. No ring forms.

  Tube 2. Antigens are present but no antibodies are present to react with them. No ring forms.

  Tube 3. Antigens against BSA are present but normal rabbit serum does not contain antibodies against BSA. No ring forms.

Tube 4. Antigens against BSA are present but normal rabbit serum does not contain antibodies against BSA. No ring forms.

Tube 5. Both antigen and antibodies are present. A ring forms.

e. The precipitin ring test was originally developed for forensics work. By using blood serum from the patient to be tested, antigens for the disease can be introduced. If the patient possesses the antibodies to the antigens, a ring will form, indicating active immunity.

f. A person with type B blood has antibody A in the blood plasma. If the person receives type A blood by mistake, the antibody-antigen reaction causes agglutination of the red blood cells. They would clump together, clogging capillaries and smaller vessels and would impede circulation. This agglutination is similar to the antigen-antibody reaction that causes the ring to form in the precipitin ring test.

Laboratory Review Questions and Problems

1.

Formation of:	Blastula	Early gastrula	Late gastrula	Neurula
Processes occurring	cleavage (cell proliferation)	gastrulation (migration and invagination)	gastrulation (determination, cell differentiation)	neurulation (induction, organ differentiation)
Type of structure formed	blastula (blastocoel)	gastrula (blastopore, archenteron)	gastrula	neurula (brain and spinal column)
Characteristics of structure	hollow sphere	blastopore	archenteron	neural tube
Significance of stage	cell positions impart prospective significance	differentiation of germ layers	cells are determined	tissues are competent to be induced (to form organs)

2.

	Sea urchin	Frog	Chick
Type of egg	microlecithal, isolecithal	mesolecithal, moderately telolecithal	macrolecithal, extremely teloleccithal
Type/pattern of cleavage	holoblastic, equal	holoblastic, unequal	meroblastic
Distinguishing characteristics of blastula	uniform, hollow sphere	blastocoel displaced to the animal pole	very small blastodisc on a large quantity of yolk
How gastrulation occurs	invagination	migration	migration
Events of neurulation	no neurulation; embryo develops into adult	indirect development; tadpole develops into adult	direct development; embryo develops directly into adult

3. Cytoplasmic determinants affecting early development (up to gastrula) are proteins made from stored messenger RNA. These stored mRNAs were produced by the maternal genome and laid down in the developing egg's cytoplasm. The orientation of early cleavage planes has been demonstrated to be affected by the presence of cytoplasmic determinants. For instance, during spiral cleavage, whether the spiraling occurs to the left or right is controlled by cytoplasmic determinants genetically corresponding to the maternal genotype.

4. Cytoplasmic determinants (see 3 above) are an example of materials laid down by the maternal genome and added to the developing egg before meiosis eliminates one set of homologous chromosomes. (Recall that the egg does NOT have the same genetic content as the 2*n* adult.) Thus, it is possible for stored mRNA to represent products of genes no longer present in the egg (and maternal genes no longer present in the offspring).

5. Homeotic genes represent a series of genes that control the development of segmentation in organisms. Each homeotic gene contains a conserved homeobox sequence that codes for a DNA regulatory protein that can turn on a cascade of other genes responsible for the development of segmental patterns.

6. The heavy chain is composed of a constant (C) segment and a variable segment that is composed of V, D, and J segments. The constant region (selected from a family of 8 segments) is next to the J segment and determines the type of heavy chain. For instahce, a Cμ segment next to a J segment is typical for a IgM antibody (see Figure 19H-1, p. 19-17). Later in life, a different segment may be inserted next to the J segment (class switching) and will result in the antibody now falling into one of the other immunoglobulin classes (IgA, IgD, IgE, or IqG).

7. Variability among immunoglobulins is the result of splicing pieces of DNA together to form different combinations in transcripts (which then can be further modified by removal of introns).

8. Antigens are foreign molecules, usually proteins, that are capable of eliciting the immune response. Plasma cells, a type of β lymphocyte, secrete antibodies (immunoglobulins) in response to antigens.

9. If an epidemic is spreading rapidly, it would be advantageous to conduct a simple precipitin ring test that would provide information about the proportion of the population that is immune. This information would help to "manage" a large scale epidemic.

10. Amnion—   filled with amniotic fluid to protect the embryo
    Chorion—   protection and gas exchange (with the vessels of the allantois)
    Allantois—   waste storage and gas exchange
    Yolk sac—   vitelline vessels associated with the yolk sac deliver yolk nutrients to the embryo

11. Vitelline vessels in the chick embryo are found on the surface of the yolk sac. Blood flows to the yolk sac via the vitelline arteries branching from two large omphalomesenteric arteries. The vitelline arteries ramify into many small vessels. In early stages of embryonic development, these vessels assist with oxygenation of the blood, picking up oxygen from the vast vessel surface area exposed on the yolk sac surface (oxygen enters the egg shell by diffusion). At a later stage, the vessels pick up nutrients from the yolk sac. The yolk is made soluble by digestive enzymes produced by the endodermal cells of the yolk sac. The nutrient-

laden blood is returned to the sinus venosus of the heart.

12. The human placenta is responsible for oxygenation of fetal blood and for removal of wastes and carbon dioxide. In the chick, the chorioallantoic membrane (formed by the fusion of the chorion with the allanois) system carries out the same functions. In both the chick and human, the embryo is surrounded by a fluid-filled protective amnion.

13. The allantois is not used because waste disposal is taken car of by the placenta.

More:

Laboratory 20

The Genetic Basis of Evolution I—Populations

EXERCISE A   Understanding Variation

PART 1   Measuring Cephalic Index

p. 20-2   a. Male average: 80.0 ± 1.7. Female average: 80.3 ± 2.4.
The class average cephalic index will be somewhere in the range from 70 to 90 (A. J. Kelso, 1970. *Physical Anthropology*). A low cephalic index indicates a narrow head and a high cephalic index indicates a broad head.

b. The range of measurements should be greater for females. There should be more variation in cephalic index among the females. A simple way to arrive at this conclusion is to compare the standard deviations of cephalic index for each sex.

p. 20-3   **Figure 20A-2** *Bar graph for class cephalic index data*

					82.7
		82.1			84.5
		80.3			83.0
		81.9			82.9
	78.9	82.0			82.6
	76.2	82.6			82.3
	77.5	80.1		79.4	81.9
	77.8	80.3		79.0	81.9
	78.6	80.5		79.0	81.5
	78.9	80.8		78.7	81.5
	79.4	80.9		78.6	81.5
	79.4	81.3		78.5	81.3
	79.7	81.4		77.9	80.8
	79.4	81.5		76.8	80.8
	77.0	81.6	74.6	76.2	80.1
70-74.9	75-79.9	80-84.9	70-74.9	75-79.9	80-84.9
MALES			**FEMALES**		
average: 80.0 ± 1.7			**average: 80.3 ± 2.4**		

<50	50-51	52-53	54-55	>55	<50	50-51	52-53	54-55	>55
		52					53		
		53					53		
	51	52					53		
	51	52					53		
	51	52					53	55	
	51	53					53	55	
	51	53					53	54	
	51	52					53	54	
	51	53				51	52	54	
	51	52	55			51	52	54	
	50	52	55			51	52	54	
48	50	52	54			50	52	54	56
<50	50-51	52-53	54-55	>55	<50	50-51	52-53	54-55	>55
MALES					FEMALES				
average: 51.8 ± 1.5					average: 53.0 ± 1.4				

p. 20-3   a. Male average: 51.8 ± 1.5. Female average: 53.0 ± 1.4
          b. The ranges of average relative sitting heights for females and males are approximately equal.

p. 20-4   c. The variation is average sitting heights is similar in males and in females. A simple way to arrive at this conclusion is to compare the standard deviations of relative siting height for each sex.
          d. The variation in relative sitting height is continuous.
          e. A single gene may control a discontinuous (discrete) trait such as corn seed color, but multiple genes are probably involved in the control of discontinuous traits such as relative sitting height and cephalic index.

EXERCISE B   Estimating Allelic Frequency from a Population Sample

p. 20-4   a. $p = 0.35$.

p. 20-5   b. Yes. Population 3 ($p^2 = 0.25$, $p = 0.05$; $q^2 = 0.25$, $q = 0.05$; $2\,pq = 2 \times 0.05 \times 0.05 = 0.50$)

p. 20-6   c. No. $p = (1.0 \times 0.1) + (0.5 \times .45) + (0 \times .45) = 0.1 + 0.0225 = 0.1225$ and $q = 1 - 0.1225 = 0.8775$. Comparing observed and expected—

	AA	Aa	aa
Observed	.10	.45	.45
Expected	.015	.215	.770

it is evident that the population is not in equilibrium

p. 20-7   a. $2pq$ is the frequency of the heterozygote (*Aa*), which is a taster since the allele for tasting PTC is dominant.

Table 20B-1 Phenotypic Proportions of Tasters and Nontasters of PTC and Frequencies of the Determining Alleles

Sample	Phenotypes		Allele Frequency	
	Tasters ($p2 + 2pq$)	Nontasters (q^2)	p	q
Class population	0.46	0.54	0.27	0.73
North American population	0.55	0.45	0.33	0.67

p. 20-8   **b.** Class sample frequencies approximate population estimates.
   c. Potential discrepancies could be explained by variations in local populations (greater proportions of indigenous ethnic groups).

EXERCISE C   The Founder Effect as an Example of Genetic Drift

p. 20-8   Evolution will not occur under these conditions because there is no change in allelic frequencies.

p. 20-9   **a.** Yes, there are 6 cups with zero rare alleles.
   b. 20% of the populations are without the rare alleles.
   c. Yes, there are several populations containing 2 or 3 of the rare alleles. That would be an allelic frequency of $q = 0.2$ or $q = 0.3$ rather than $q = 0.1$ as in the parent population.
   d. The allelic frequency of $q = 0.1$ appears in 53% of the populations.
   e. The allelic frequencies of small founder populations can vary from that of the parent population.

p. 20-10   **Table 29C-1 Data for Comparison of Allelic Frequencies in Parent Population and Founder Populations**

Parent generation $p = 0.9$; $q= 0.1$	Number of Founder Populations (cups) Containing the Following Numbers (0-6) of Rare Alleles							
	0	1	2	3	4	5	6	
Group 1	2	9	0	2				
Group 2	3	5	2	1				
Group 3	1	5	2	1				
Group 4 .								
…								
Total no. of cups with indicated no. of rare alleles	6	16	4	4				Total cups 30
Percent of all populations with indicated no. of rare alleles	20%	53%	13%	13%				

Table 20-C-2 Range of Allelic Frequencies in Founder Populations

Population Size (number of individuals)	Lowest q	Highest q
5	0	0/10; 3/10
10	0	0/20; 3/20
25	0.08	4/50; 7/50
50	0.09	9/100; 14/100

Calculate the lowest allelic frequency by dividing by lowest number of rare alleles found in any cup by the total number of alleles (beads) for the population (cup). Example: the lowest number of rare alleles in any cup containing 10 beads (5 individuals) might be 2. Therefore, 2/10 = 0.2. Calculate the highest alleleic frequency by dividing the highest number of rare alleles found in any cup by the total number of alleles (beads) for that population (cup).

p. 20-11 **Table 20C-3 Effect of the Size of Founder Populations (5, 10, or 25) on the Frequency of Rare Alleles**

	Number of Founder Populations with 0 Rare Alleles for Populations of 5, 10, or 25 Individuals		
	5	10	25
Group 1	2	0	0
Group 2	3	1	0
Group 3	1	1	0
...			
Total no. of cups or groups of cups with zero rare alleles	6	2	0
Percent of all populations with zero rare alleles	20%	13%	0%

7. Class data (population size = 10, 2 cups per population)

	Number of Rare Alleles								
	0	1	2	3	4	5	6	7	8
# cups	2	2	6	2	1	1	1	0	0
%	13	13	40	13	7	7	7	0	0

f. Yes. Two populations have 0 rare alleles.
g. 13% of the populations have zero rare alleles. This is less than when 5 individuals were considered.

10. Class data (population size = 25, 5 cups per population)

	Number of Rare Alleles								
	0	1	2	3	4	5	6	7	8
# cups	0	0	0	0	1	1	1	3	0
%	0	0	0	0	17	17	17	50	0

p. 20-12 **h.** Now, 0% of the populations have 0 rare alleles. This is less than when 10 individuals were considered.

12. Class data (population size 50, all cups)

	Number of Rare Alleles								
	0	**...**	**8**	**9**	**10**	**11**	**12**	**13**	**14**
# cups	0	0	0	1	0	0	0	1	1
%	0	0	0	33	0	0	0	33	33

For a population of 50 individuals containing 9 recessive alleles:
A $(p) = 0.91$
a $(q) = 0.09$
For a population of 50 individuals containing 13 rare alleles:
A $(p) = 0.86$
a $(q) = 0.14$

i. When populations are very small (5 or 10 individuals) there is the likelihood that some populations will not contain the rare allele. However, it is also possible for the allelic frequency, by chance, to be higher than the parent population ($p = 0.9$ and $q = 0.1$) as seen by populations with 3 out of 10 alleles ($p = 0.7$, $q = 0.3$) being rare.

j. The smaller the founder population, the more likely that the frequencies of alleles will vary from the parent population. As the population gets larger (for example 50 individuals), the allelic frequencies $p = 0.86$ and $q = 0.14$, the closer you get to an allelic frequency typical of the parent population. Notice that the percentage of populations with zero alleles decreases as the population gets larger so the rare allele will not become extinct (fixed) as quickly.

k. Drift for a small population could lead to the loss of fixation of an allele much more quickly than possible in larger populations.

EXERCISE D   The Role of Gene Flow in Similarity Between Two Populations

p. 20-13 a. Each transfer represents a removal of alleles from one population and the addition of those alleles to the other.

Table 20D-1 Changes in Allele Frequencies of Two Populations Experiencing Gene Flow

Round	Tray 1		Tray 2		Differences Between Trays	
	p_1 (0.70)	q_1 (0.30)	p_2 (0.90)	q_2 (0.10)	$p_1 - p_2$ (0.20)	$q_1 q_2$ (0.20)
One	0.76	0.24	0.84	0.16	0.08	0.08
Two	0.72	0.28	0.88	0.12	0.16	0.16
Three	0.80	0.20	0.80	0.20	0	0
Four	0.84	0.16	0.76	0.24	0.08	0.08
Five	0.86	0.14	0.80	.020	0.06	0.06

b. The difference in the final frequency of allele A ($p_1 - p_2$) should be less than the difference in the original frequency as populations approach equality due to continued gene flow. This is the case, changing from a difference of 0.20 to a difference of 0.06.

p. 20-14 c. Allele frequencies should show the greatest change in the early rounds, when the populations are most different.

    d. Genetic makeup should trend towards similarity. Therefore, after 10 or 20 rounds populations should be approximately equal.

    e. Unimpeded gene flow increases the similarity of originally dissimilar neighboring populations.

    f. Transferring only 5 beads simulates a situation in which the rate of gene flow is reduced. Therefore, the populations will require more generations to achieve similarity. If gene flow is unidirectional (only from tray 1 to tray 2), the absolute size of the tray 1 population will decrease and the rare allele will trend towards extinction in the tray 1 population.

EXERCISE E   The Effect of Selection on the Loss of an Allele from a Population

PART 1   The Effects of Recessiveness on Deleterious Alleles

A   Allele *B* Does Not Mask Allele *b*

p. 20-15   **a.** Twenty beads represent only 10 individuals because each individual has 2 alleles for every trait..

p. 20-16   **Table 20E-1   Effects of Recessiveness on the Persistence of a Deleterious Allele When *B* Does Not Mask *b***

Generation	Frequency of Alleles		Frequency of Genotypes		
	p (*B*)	q (*b*)	p^2 (*BB*)	$2pq$ (*Bb*)	q^2 (*bb*)
Parent	0.8	0.2	0.64	0.32	0.04
F_1	0.89	0.11	0.78	0.22	0
F_2	0.95	0.05	0.92	0.06	0.02

p. 20-17   **b.** The frequency of the dominant allele has increased slightly while the frequency of the recessive allele has decreased.

    c. The genotypic frequencies of the homozygous dominant are increasing while those of the homozygous recessive rapidly change toward zero. Since, in the heterozygote, *B* does not mask *b*, the effects of *b* are felt and the frequency of heterozygotes is also decreasing rather rapidly.

    d. The allelic frequency of the recessive allele continues to decrease. The genotypic frequency of the homozygous recessive actually increases a bit. Because of the recessive allele refuge, it is possible for the few recessive alleles still existing in the fewer and fewer heterozygotes to actually be combined, by chance, in a homozygous recessive individual.

    e. This trend occurs because the recessive allele is lethal. As it gets more rare, however, it can "hide" in the very small heterozygous population.

    f. Data are representative of the class data although the specific numbers vary. All dominant allele frequencies increase while recessive allele frequencies decrease.

B   Allele *B* Masks Allele *b*

p. 20-18   Table 20E-2 Effects of Recessiveness on the Persistence of a Deleterious Allele when *B* Masks *b*

Generation	Frequency of Alleles		Frequency of Genotypes		
	p (*B*)	q (*b*)	p^2 (*BB*)	$2pq$ (*Bb*)	q^2 (*bb*)
Parent	0.8	0.2	0.65	0.30	0.05
F_1	0.89	0.11	0.75	0.25	0
F_2	0.86	0.14	0.77	0.19	0.04

g. The allele frequency of the dominant allele is increasing but not as fast as when "*B* does not mask *b*." Also, the allele frequency of the recessive allele is decreasing but not as rapidly as in "*B* does not mask *b*."

h. This occurs because *B* masks *b* so that more heterozygotes survive. Thus, more recessive alleles are maintained in the population. This allows more heterozygous individuals to remain and gives a greater opportunity for homozygous recessive individuals to be formed (although they do not survive).

i. Masked recessive alleles persist longer than those that are not masked because they "hide" in the heterozygotes without putting the heterozygotes at a disadvantage. For this reason, heterozygotes can live to maturity and reproduce, resulting in more heterozygotes and even homozygotes.

j. Yes, *b* persisted longer when masked.

k. Recessiveness would hinder the spread of beneficial alleles because they would probably not be expressed in the heterozygote (unless penetrant or incomplete dominance exists).

PART 2   Heterozygote Advantage

p. 20-18   When matings between *Bb* and *Bb* occur, you expect ¼ *BB*, ½ *Bb*, and ¼ *bb*. In this case, the frequency of *B* = 0.5 and the frequency of *b* = 0.5.

p. 20-19   Table 20E-3 Effects of Heterozygote Advantage on Allelic Frequencies

Generation	Frequency of Alleles		Frequency of Genotypes		
	p (*B*)	q (*b*)	p^2 (*BB*)	$2pq$ (*Bb*)	q^2 (*bb*)
Parent	0.8	0.2	0.650.7	0.2	0.1
F_1	0.38	0.62	0	0.75	0.25
F_2	0.42	0.58	0.08	0.76	0.16

a. The allelic population of F_1 and F_2 approach frequencies of $p = 0.5$ and $q = 0.5$ which would be expected if a population increases in the number of heterozygotes. (A population of all heterozygotes *Bb* × *Bb* gas frequencies of $p = 0.5$ and $q = 0.5$.)

b. No, the recessive allele never disappears because it is part of the heterozygote.

c. p and q are moving toward $p = 0.5$ and $q = 0.5$ because, in a total heterozygote population, *Bb* × *Bb* gives ¼ *BB*, ½ *Bb*, and ¼ *bb* so that $p = 0.5$ and $q = 0.5$.

p. 20-20 d. No. The maximum would be $p = 0.5$ and $q = 0.5$ when the entire population is composed of heterozygotes only.

e. Genotypic frequencies show a decrease in homozygous dominants while homozygous recessives remain fairly steady (although one would expect the number of homozygous recessive individuals to also decrease). The heterozygote helps to maintain the recessive allele in numbers that are even higher than what the population started with so that more homozygous individuals can be formed (although they die before reproducing).

f. Yes, numbers are typical of the class data.

g. The genotype of heterozygotes contains both the dominant and recessive alleles, and since this is the favored genotype, both alleles are maintained in the population.

h. Once the heterozygous advantage disappears, the deleterious recessive allele will begin to disappear because the number of heterozygotes will decrease. The deleterious allelic frequency will decrease rapidly if not masked by the dominant, or more slowly if masked by the dominant.

i. The *b* allele persists longer in this exercise because the number of heterozygotes increases under heterozygote advantage conditions. Since each heterozygote includes a recessive allele, the recessive allele persists longer. When *B* masks *b*, the effects of the recessive allele may be masked, but there is no advantage to being heterozygous, so the number of heterozygotes (and thus the frequency of the recessive allele) will not increase.

Extending Your Investigation: Selection Pressure

p. 20-20   HYPOTHESIS: The more conspicuous a prey is, the more frequently it will be captured and selected against.

NULL HYPOTHESIS: The color of prey (yarn) has no effect on selection.

Prediction—the more conspicuous "wooly worms" (colors contrasting with the background) will be preferentially removed.

Independent variable—phenotype of prey (color).

Dependent variable—number of prey captured.

p. 20-21   **Table 20-I Chi-Square Calculations**

Color	A No. Observed (collected)	B No. Expected (by chance)	C Observed − Expected	D (Observed − Expected)^2	E (Observed − Expected)^2/ Expected
1. white	16	13.88	2.12	4.49	0.323
2. green	18	13.88	4.12	16.97	1.223
3. pink	19	13.88	5.12	26.21	1.888
4. blue	17	13.88	3.12	8.71	0.701
5. beige	12	13.88	−1.88	3.53	0.254
6. brown	5	13.88	−8.88	78.85	5.680
7. black	6	13.88	−7.88	62.09	4.473
8. red	18	13.88	4.12	16.97	1.223
...					
Total number observed = 111			Sum of chi-square values = 15.765		

p. 20-22   Results—support the hypothesis. There is selection based on color.

Results allow the null hypothesis to be rejected.

The prediction was correct—brighter colors are selected against.

Conclusion—phenotypes can affect selection.

Only black, brown, and beige were subjected to positive selection pressure.

Because they were not selected by the predator, their genes remain in the gene pool.

The gene frequencies of black, brown, and beige will increase while the gene frequencies for all other colors will decrease.

Increase the number of trials if the population is large (infinite). However, as the more conspicuous pieces of yarn are removed from a finite population, more of the less conspicuous pieces of yarn will be sampled. In this case, the population size may need to be increased (more pieces of yarn scattered over a wider area).

Laboratory Review Questions and Problems

1. An allelic frequency (i.e., A) is the fraction of all alleles at a particular (A/a) locus that are A.

2. You can always calculate allelic frequencies from a complete list of genotypic frequencies whether or not populations are in Hardy-Weinberg equilibrium.

3. A or $p = 0.2$
   a or $q = 0.8$

4. The expected genotypic frequencies if the population is in Hardy-Weinberg equilibrium would be
   $$AA = p^2 = 0.04$$
   $$Aa = 2pg = 0.32$$
   $$aa = q^2 = 0.64$$
   Since the expected and actual genotypes do not agree, the population is NOT in Hardy-Weinberg equilibrium.

5. If observed and expected genotype frequencies are equal, then the population is at Hardy-Weinberg equilibrium.

6. $p^2 = 0.49$ (frequency of the homozygous dominant, DD)
   Therefore $p = 0.7$
   Since: $p + q = 1$, then: $q = 0.3$
   Therefore: $q^2 = 0.09$ (frequency of the homozygous recessive, dd)
   $\qquad$ $2pq = 0.42$ (frequency of the heterozygote, Dd)
   Therefore: the frequency of $D = p = 0.7$
   Therefore: the frequency of $d = q = 0.3$

7. You can only calculate genotypic frequencies from allelic frequencies if you assume Hardy-Weinberg equilibrium. Since the class population is small and violates the size conditions of the Hardy-Weinberg hypothesis, you cannot assume Hardy-Weinberg equilibrium and should not try to calculate genotypic frequencies.

8. You can calculate genotypic frequencies from allelic frequencies ONLY when populations are in Hardy-Weinberg equilibrium.

9. You can estimate allelic and genotypic frequencies from the fraction of homozygous recessives in a population ONLY if you assume Hardy-Weinberg equilibrium. In question 6 you CAN estimate genotypic frequencies because the population IS in Hardy-Weinberg equilibrium.

10. The frequency of $d = 0.4 = q$
    Therefore, $q^2 = 0.16$ (frequency of the homozygous recessive, dd)
    $p + q = 1, p + 0.4 = 1$
    Therefore: $p = 0.6, p^2 = 0.36$ (frequency of the heterozygote, Dd)
    $2pq = 2 (0.6) (0.4) = 0.48$ (frequency of the heterozygote, Dd)

11. The conditions necessary for the Hardy-Weinberg equation to predict allele frequencies are: (1) No mutation of alleles. (2) Mating must be random. (3) The population must be-large. (4) No immigration nor emigration can occur. (5) There can be no selective advantage for a particular genotype. Thus, the Hardy-Weinberg equation is a "null" hypothesis because it is indicative of the absence of evolutionary change.

12. The allelic frequency probably will change.

13. Drift is a change in allelic frequencies that results from random mating. Drift is usually exaggerated in small populations because the gene pool may differ significantly from the parent stock.

14. Non-random mating favors certain phenotypes and thus certain genotypes. Migration brings new alleles into a population (immigration) or causes them to leave a population (emigration), changing allelic frequencies in the population.

15. Yes. Evolution is a change in allele frequencies over time. Drift can change allele frequencies, especially in small populations.

16. Exercise B, # 3. Population is not large.
    Exercise C, # 3. Population is not large.
    Exercise D, # 4. Migration takes place.
    Exercise E, # 5. Selection takes place.

17. Part 1 A. *BB—b* is deleterious and is not masked by *b*.
    Part 1B. *BB, Bb—b* is deleterious but is masked by *B*.
    Part 2. *Bb*—other genotypes are selected against.

18. Solving.this problem is not appropriate unless you are assured that the population in question is in Hardy-Weinberg equilibrium.

Laboratory 21

Genetic Basis of Evolution II—Diversity

EXERCISE A	Understanding Evolutionary Classification

p. 21-2   1. Kingdom Animalia
        Phylum Chordata
            Subphylum Vertebrate
                Class Mammalia
                    Subclass Eutheria
                        Order Primates
                            Family Hominidae
                              Genus *Homo*
                                Species *Homo sapiens*

2. Monophyletic groups in Figure 21A-1 (p. 21-3) include each species cluster (species B, C and D, F, G and H, and I, J and K) and the entire phylogram if the ancestral species A is included.

3. There are no polyphyletic groups in the phylogram as drawn. To be polyphyletic, a group would have to include species with different ancestors—for example, if species G were included with species C and D in one group, it would be polyphyletic.

4. If species C through H and their ancestors (which are common to species J and K) are grouped together separately from species J and K and their immediate ancestor, species I, they would form two paraphyletic groups. Likewise, if G through K are grouped separately from B, C, and D and their common ancestor (A), they would be considered paraphyletic.

3. In Figure 21A-2 (p. 21-3), the entire primate order is believed to be monophyletic. Old World and New World monkeys may be polyphyletic if considered with lemurs and tarsiers but the diagram suggests a common origin and monophyly. If Old World and New World monkeys, as well as their common ancestor are considered, this would b a paraphyletic taxon as related to the rest of the primates. If, for some reason, the current Family Hominidae (humans) were thought to be distinct enough to be considered an order instead of a family, then they and the remainder of the Primates would be paraphyletic.

p. 21-2,3   6. <u>Cartilaginous fishes</u>—Class Chondrichthyes
      cartilaginous skeleton (an embryonic condition useful in reducing density); no lungs or swim bladders
      <u>Bony fishes</u>—Class Osteichthyes
      bony skeletons, dermal scales, lungs or swim bladders
      Note: Fossil agnatha (jawless fishes) first developed bony skeletons and survive as highly specialized lampreys and hagfishes). Other fishes and tetrapods have jaws. The extinct placoderms constituted a diverse class that experimented with appendages, scales, and even appear to have developed a lung or swim

bladder. Among bony fishes, Class Osteichthyes, there are two relatively different lines (the Actinopterygii, ray-finned fishes, which constitute the bulk of living fishes; and the Sarcopterygii, lobe-finned fishes, represented today by only four surviving species, the lungfishes and coelocanth). The Sarcopterygii have lungs and the bones of their appendages are homologous with those of tetrapods. These relationships are explored in Laboratory 27 (Diversity—Chordata).

Amphibians—Class Amphibia
two pairs of appendages, gills in larvae and lungs (in most) adults, moist skin, unprotected eggs (laid in water or moist environments)

Reptiles—Class Reptilia
first class with closed (cleidoic), amniotic egg, epidermal scales predominate

Note: all organisms to this point are ectothermic—dependent on their environment (in some measure) for their body temperature (and activity). Birds and mammals are endothermic—body temperature is maintained at nearly constant (and high) levels by physiological means giving them access to colder environments but requiring a higher level of energy expenditure to maintain optimal body temperature.

Birds—Class Aves
forelimbs modified as wings; feathers used for flight and as insulating structures

Mammals—Class Mammalia
hair used as insulating structure; most give birth to living young (viviparity) and nourish them with maternal secretions (milk, produced in mammary glands), young have an extended period of dependence on parents

a. All vertebrates have the common chordate characteristics (notochord, dorsal tubular nerve cord, pharyngeal pouches, post-anal tail) plus vertebrae, a characteristic heart structure with at least three chambers, a liver, etc.

b. Each class has distinguishing characteristics. Trends include the evolution of jaws, separating jawless fishes from all other vertebrates; the evolution of paired appendages early in ancestral fishes; the appearance of a lung/swim bladder that could develop into a functional respiratory organ in the line leading to tetrapods and a hydrostatic organ in the ray-finned bony fishes, a progression from heavy dermal armor (scales) in fishes to epidermal coverings that retard desiccation (and give rise to feathers and hair in birds and mammals respectively); ectothermy vs. endothermy; development of the amniotic egg; oviparity vs. viviparity (developed in several lines besides mammals), etc.

c. Homologies (relationships by descent) can be traced through the evolution of all vertebrate stuctures. From common methods of cleavage and development of the blastopore (vertebrates are deuterostomes; the blastopore is associated with the caudal opening of the gut or enteron), to features of morphogenesis, embryogenesis, and development of systems, derivations and adaptations can be discerned. For example, pharyngeal pouches become subdivided in vertebrate ancestors to form pharyngeal slits used in filtering food from the surrounding water. This feeding circulation brings oxygenated water into the vicinity of the pharyngeal arches separating the slits. With the addition of a vascular bed (to form gills), this structure becomes associated with gas exchange and, as jaws appear, looses its feeding function. With the evolution of lungs (and the transition to land), respiratory functions are lost but various elements give rise to structures in the tetrapod body—an anchor for the base of the tongue, bones of the middle ear, the larynx and associated structures, and a variety of glands (tonsils, parathyroids, thymus, etc.). Lungs develop as an evagination from the floor of the pharynx that, in higher bony fishes, looses its pharyngeal connection and becomes a swim bladder, a hydrostatic organ, rather than a respiratory organ. A variety of themes can be elaborated to answer this question.

p. 21-4 **d.** Structures that are similar in function but not related by ancestry are often said to be homoplastic. Gills and lungs are both respiratory organs and develop from pharyngeal endoderm, but the former are associated with the pharyngeal pouches and the latter with a diverticulum from the floor of the pharynx suggesting very different origins. All vertebrates are covered by skin but in fishes, dermal, often bony, scales predominate while in tetrapods dermal bone is largely lost and the surface is protected by cornified (keratinized) epidermal structures. The patterns of skeletal support in the fins of most bony fishes differs from these patterns in the lobe-finned fishes and tetrapods.

8. **Cladogram**

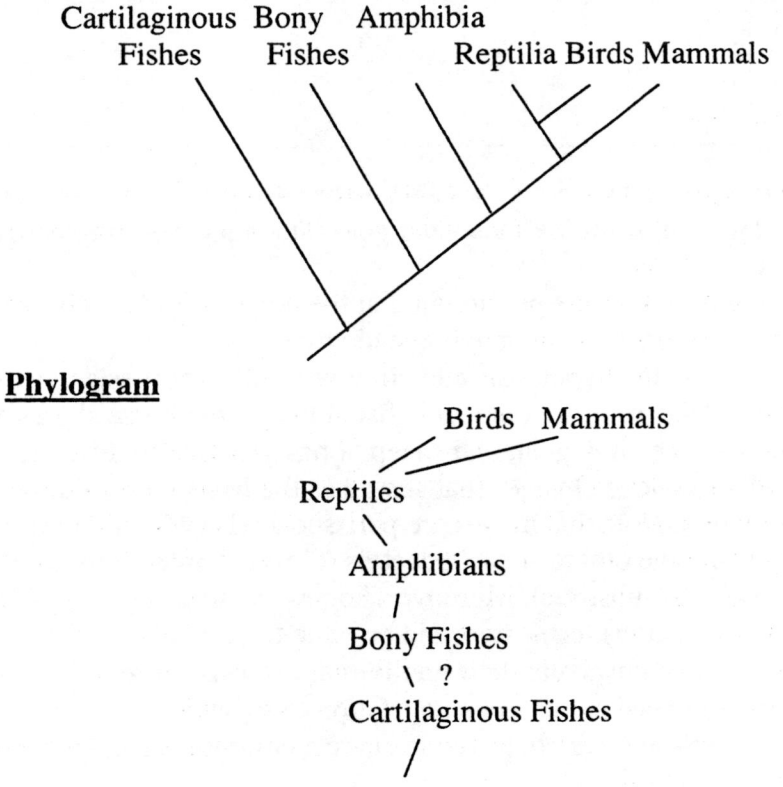

Phylogram

see the expanded phylogram Laboratory 27, Figure 27-II

p. 21-8   HYPOTHESIS: Electrophoretic patterns for LDH isozymes of the artiodactyls, the sheep, goat, and cow, will be closer to each other than any will be to the patterns for the horse.

NULL HYPOTHESIS: There will be no difference among the four species with respect to their electrophoretic patterns for LDH isozmes.

Prediction—the sheep, goat, and cow will show similar patterns to each other and will differ from patterns shown by the horse.

Independent variable—type of animal.

Dependent variable—banding pattern.

p. 21-8   LDH banding patterns:

LDH-3 LDH-2 LDH-1

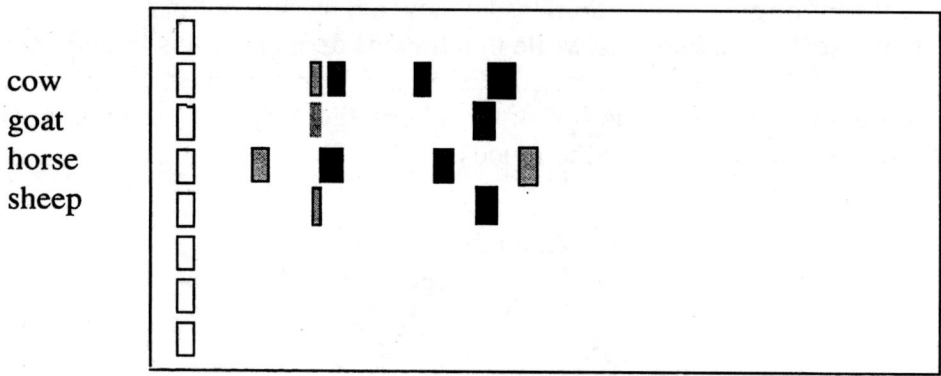

cow
goat
horse
sheep

a. The cow, goat, and sheep show LDH patterns that are most similar. The cow is slightly different from the sheep and goat. The horse banding pattern is very different.

b. Yes, this agrees with the prediction that the horse would be different from the rest, based on its position on the phylogenetic tree.

Results support the hypothesis and allow the null hypothesis to be rejected. From this experiment, we conclude that although the horse shares a common ancestor with the cow, goat, and sheep, it has genetically different LDH isozymes as a result of genetic changes that served as the basis for evolution to produce a separate group. Note that horses are perissodactyls (with a single toe touching the surface) while the others are artiodactyls (cloven-hoofed animals with two toes in contact with the substrate). Moreover, horses are specialized as hind-gut (large intestine and cecum) fermenters and are able to survive on drier and more woody vegetation while cows and their relatives are fore-gut fermenters (with modified stomachs) and need some green vegetation to survive.

c. Monophyletic—they are believed to share a common ancestor found among the primitive ungulates.

Laboratory Review Questions and Problems

1. Phylogeny—the lines of descent among a group of organisms including the common ancestor and the degree of divergence of the various branches.

   Taxon—a level or category used in classification; e.g., phylum, order, family, species.

   Binomial nomenclature—first applied consistently by Carolus Linnaeus; every species is identified by a unique species name consisting of a genus and a specific epithet, e.g., *Homo sapiens*.

   Monophyletic taxon—any taxon containing all known descendants from a common ancestor and no other species in any other group.

   Polyphyletic taxon—any taxon that contains representatives that do not share the same nearest common ancestor. With convergent or parallel evolution, ancestry may be difficult to unravel and lack of information may produce polyphyletic groups. This is particularly true for taxa that are less well known.

   Paraphyletic taxon—a taxon that does not contain all the derivatives of that group. We know that the ancestry of both birds and mammals can be traced to different lines of reptiles, but most biologists recognize these three groups as separate classes, recognizing the many advanced features in each.

   Cladistics—a school of systematics that seeks to determine branches among monophyletic lines.

   Evolutionary systematics—the practice of arranging taxa in monophyletic assemblages based on their degree of difference as well as their inferred genealogy.

   Homology—relationships among structures based on ancestry and descent; relationships with a genetic basis.

   Analogy—relationships among structures based on resemblance. The wings of insects and flying vertebrates are analogous. Among pterosaurs, birds, and bats wings are derived from the forelimb and are homologous to that extent. However, in pterosaurs, the flight surface was formed by a membrane essentially supported by only one finger, in birds, the hand is reduced and supports non-living feathers that form the flight surface, while in bats several fingers support the membrane of skin used in flight. Thus, there are hierarchies of relationships that need to carefully analyzed.

   Homoplasy—relationships among structures based on similarity of function. Gills and lungs are homoplastic because both are involved in respiration but they are not homologous (related by descent) nor analogous (of similar appearance).

   Phylogeny—relation by descent. Phylogeny is the study of the evolutionary history and present interrelations among various taxa.

   Classification—the linear ordering of taxa to produce a useful scheme for storing and retrieving information, arranging specimens, and directing learning. Note that phylogenies are branching, not linear. Classifications usually list all members of "less advanced" groups first, moving to more derived groups, and will include both extinct and living taxa arranged hierarchically.

   Systematics—the science of determining evolutionary relationships among organisms (taxonomy is a branch of systematics concerned with identifying and naming organisms).

2. a. B and D are most closely related.
    b. C is least related to the others.
    c.

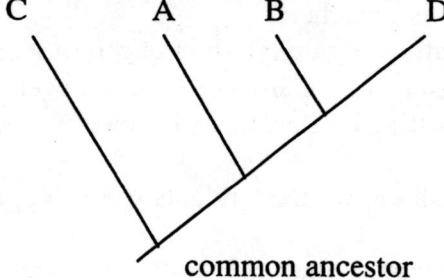

common ancestor

Laboratory 22

Diversity—Kingdoms Eubacteria, Archaebacteria, and Protista

PART I DOMAINS BACTERIA AND ARCHAEA

EXERCISE A Morphology of Bacteria

p. 22-3
a. Round.
b. In chains.
c. A. bacillus: *Escherichia coli* (single); *Bacillus subtilis* (chains)
   B. coccus: *Staphylococcus aureus* (clusters)
   C. spirillum: *Spirillum rubrum* (single)
2.

Color	Gram + or –
A. pink	–
B. purple	+
C. pink	–

d. In general, cocci bacteria are gram positive with a few exceptions such as the meningococci, gonococci, and catarrhalis groups.
e. Purple.
f. Yes. Gram + bacteria have thick peptidoglycan (murein) cell walls. If they are unable to synthesize this material, they will be unable to grow and reproduce.

p. 22-4
5. Plates will usually show whitish, wet-looking bacterial growth, some yellow colonies, and usually green mold or whitish fungal growth. Results are variable.

EXERCISE B Characteristics of Bacteria: Sensitivity to Antibiotics

p. 22-5 Table 22B-1 Effects of Antibiotics

Antibiotic	Effect on Growth	
	Plate 1: Gram –	Plate 2: Gram +
A. Erythromycin	–	+
B. Tetracycline	+	+
C. Ampicillin	+	++
D. Gentamycin	+	–
E. Penicillin	–	++

Table 22B-2 Spectrum of Antibody Activity

Antibiotic	Effective against	Spectrum
A. Erythromycin	Gram +	narrow
B. Tetracycline	Gram +, Gram –	broad
C. Ampicillin	Gram +, Gram –	broad
D. Gentamycin	Gram –	narrow
E. Penicillin	Gram +	narrow

a. A doctor would prescribe an antibiotic with a narrow spectrum for cases in which the bacterial type is known. A doctor would prescribe an antibiotic with a broad spectrum for cases in which the type of bacterial causing the infection has not been identified or if there is a possibility of multiple infections.

Extending Your Investigation: How Effective Is Your Soap?

p. 22-6 HYPOTHESIS: If bacteria are the source of perspiration odor, then deodorant soap should be more effective than regular soap in inhibiting bacterial growth.
NULL HYPOTHESIS: All soaps are equally effective against bacteria.
Prediction—the zone of inhibition will be larger for deodorant soap.
Independent variable—regular soap vs. deodorant soap.
Dependent variable—zone of inhibition.

p. 22-7 RESULTS: Plates were inoculated with *Escherichia coli*.

Test Substance	Zone of Inhibition (Diameter)
Dial soap	2 mm
Dove soap	none
Ivory soap	1 mm
Irish Spring soap	3 mm

Results support the hypothesis.
Results allow the null hypothesis to be rejected.
Conclusion—deodorant soaps appeared to be more effective against *E. coli*, a common type of bacteria found on human skin.

EXERCISE C   Nitrogen-Fixing Bacteria

p. 22-8 **a.** Plants inoculated with *Rhizobium* are more robust.
b. Plants inoculated with *Rhizobium* have root nodules.
c. Legumes
d. Cocci

EXERCISE D   Bioluminescent Bacteria

p. 22-8 **a.** The culture is luminescent. Color is yellow-green.
b. Shaking the tube aerates (oxygenates) the culture. The reaction of luciferin with FMN requires oxygen.
c. The bioluminescence gradually fades as the oxygen in the water is consumed by the reaction.
d. The bottom of the culture darkens first. The bottom has less access to atmospheric oxygen.

e. The oxygen supply in the seas is normally maintained by algal photosynthesis as well as water movements (turbulence, currents).

EXERCISE E   Diversity and Structure of Cyanobacteria

p. 22-10   **a.** Some cyanobacteria are round (*Nostoc*), while others are cuboidal-looking (*Anabaena*). Others are rectangular (*Cylindrospermum, Oscillatoria*). Some cyanobacteria have a mucilaginous sheath or coating that surrounds groups of cells either in clumps or filaments. An incomplete splitting of the cell wall keeps the cells together, forming chains.

b. Heterocysts and akinetes are common in *Nostoc, Cylindrospermum,* and *Anabaena.* Hormogonia (chains that have broken into multicellular fragments) are found in *Oscillatoria.* The environment is probably fairly anaerobic.

c. The ratio of cyanobacteria to other algae serves as an indicator of lake nutrient status. "Polluted" streams and lakes contain relatively greater biomass of cyanobacteria.

d. Cyanobacteria have no visible nucleus or other membrane-bound organelle. Differences between eukaryotes and prokaryotes are: (1) lack of membrane-bound organelles in prokaryotes; (2) cell wall composition of prokaryotes is unique (peptidoglycan); (3) genome is a single, circular molecule of double-stranded DNA in prokaryotes, whereas the eukaryote genome is organized into chromosomes; (4) prokaryote flagella are composed of flagellin protein whereas eukaryote flagella are composed of microtubules.

e. Yes. Certain antibiotics such as penicillin prevent the formation of the N-acetyl-glucosamine and N-acetyl muramic acid heteropolymers in the peptidoglycan complex (murein). If cyanobacteria are unable to synthesize murein, they cannot function properly nor can the reproduce (this requires building new cell walls).

f. No. Yes.

PART II   KINGDOM PROTISTA

EXERCISE F   Identifying Protozoans

p. 22-12   Ciliophora (*Paramecium*). Yes, ciliary movement is coordinated. Yes, contractile vacoles are visible. They fill and empty approximately every 30 seconds to 1 minute depending on environmental conditions.

p. 22-13   **a.** Organisms that lack motility must be dispersed by a vector. The mosquito is the vector for *Plasmodium.*

EXERCISE G   Symbiosis in the Termite: A Study of Flagellates

p. 22-14   **a.** Yes. There are some smaller flagellates in addition to *Trichonympha.* These include *Pyrsonympha vertens.* There are several (7-8) long flagella on each *Trichonympha.* The organisms should be placed in the Phylum Mastigophora.

b. They move erratically.

c. *Trichonympha* have special enzymes (cellulases) that are used to digest cellulose in materials such as paper or wood.

9. Symtiobic relationships such as this have evolved by chance. The termite might have ingested the protist along with other food. Those termites with populations of protists that were retained in the gut would have a selective advantage—the use of a new food source—and would have a better chance to survive.

p. 22-15 **d.** Termites and their symbionts help break down detritus to return organic nutrients to the soil so that they can be recycled through the ecosystem.

e. Bacteria share the niche as decomposers.

EXERCISE H   Primordial Slime Molds

p. 22-16 **a.** Plant-like.

p. 22-17 **b.** When the plasmodium runs out of growing space it may form a sclerotium; a dormant stage.

c. Movement is towards a food source.

d. The plasmodium continues to grow, anastomosing around the hole. The plasmodial material removed from the hole will also continue to grow.

e. Growth on the warmer side increases, due to enhanced metabolic activity.

f. *Physarum* has a negative response to light—it tends to grow away from light. Since the slime mold is heterotrophic, and as such does not contain chlorophyll, light is not needed for growth. Often, exposure to light will initiate formation of sporangia.

EXERCISE I   Water Molds

p. 22-22 **a.** Saprophytic.

EXERCISE J   Studying and Classifying Algae

EXERCISE K   Diversity Among the Green Algae: Phylum Chlorophyta

EXERCISE L   Recognizing Protists Among the Plankton

p. 22-22 **a.** Yes. Chlorophyta (desmids); Chrysophyta (diatoms). A simple taxonomic key can be used to distinguish among the different algae. Color can be used as the first characteristic used for classifying algae.

b. Yes (e.g., *Amoeba*)

c. Yes (e.g., *Paramecium*)

d. Yes, especially associated with organic debris.

e. Phytoplankton photosynthesis produces oxygen. Phytoplankton are a food source for zooplankton. Both phytoplankton and zooplankton are food sources for fish.

f. Light varies throughout the day. Subsurface light penetration and light quality vary with depth due to light absorbance by water and particles within the water. Additionally, nutrients are stratified within the water column. Therefore, phytoplankton may migrate to depths providing optimal light and nutrient conditions.

Laboratory Review Questions and Problems

1. Prokaryotes are distinguished from the eukaryotic Protista by the Eubacteria characterisitics of: (1) absence of membrane-bounded organelles, including lack of a nuclear envelope; (2) peptidoglycan cell wall; (3) circular DNA; (4) flagella composed of flagellin protein rather than the 9+2 arrangement of microtubules made from tubulin in eukaryotic protists.

2. Cyanobacteria are included with the bacteria because they are prokaryotes and have peptidoglycan cell walls encased in a mucilage coat or glycocalyx. Cyanobacteria are prokaryotes; green plants, including green algae, are eukaryotes. Cyanobacteria have peptidoglycan cell walls; green plants have cellulose cell walls. The cyanobacteria genome consists of a single, double-stranded, circular DNA molecule whereas the DNA of green plants is organized into chromosomes.

3.

Protozoan phylum	Distinguishing Characteristics	Example
Zoomastigophora	flagellar movement	*Trypanosoma*
Rhizopoda	amoeboid movement	*Amoeba*
Ciliophora	ciliary movement	*Paramecium*
Ampicomplexa	non-motile, parasitic	*Plasmodium*

4. Water molds have cellulose cell walls, rather than the chitin cell walls of most fungi. Water molds also have flagellated reproductive cells. Slime molds have a stage of the life cycle in which protist-like "myxamoeba" swarm independently before aggregating to form a (fungus-like) multicellular reproductive structure.

5.

Algal Phylum	Distinguishing Characteristics	Example
Euglenophyta	flexible, protein pellicle, rather than a cell wall	*Euglena*
Chrysophyta	silica, cellulose or $CaCO_3$ cell walls, chlorophylls a and c, fucoxanthin	Diatoms, *Dinobyron*
Pyrrophyta	cell wall composed of interlocking plates, chlorophylls a and c	*Peridinium*
Phaeophyta	multicellular, marine, chlorophylls a and c	*Laminaria* (kelp)
Rhodophyta	phycobilin pigments (red - purple), chlorophyll a	*Gracillaria*
Chlorophyta	cellulose cell walls, chlorophylls a and b	*Chlorella*

6. Euglenophytes have chloroplasts like eukaryotic algae but they also have animal-like characteristics that make them resemble the protozoans.

7. Eukaryotes and Archaeans diverged early from another line of organisms that led to the Eubacteria. Later in time, the Archaeans and Eukaryotes diverged and evolved to form remarkably different domains of organisms

Notes:

Laboratory 23

Diversity—Fungi and the Nontracheophytes

PART I   KINGDOM FUNGI

EXERCISE A   Phylum Zygomycota

p. 23-4   **a.** *n.*
b. *n; n.*
c. No.
d. Gametangia are formed by septation of the growing tips of apposing + and – hyphae.
4. Sexual reproduction is depicted by the fusing of gametangia from hyphae (1*n*) of opposite mating types and the subsequent formation and germination of a zygote (2*n*).

Extending Your Investigation: Conditions for Fungal Growth

p. 22-4   HYPOTHESIS: Molds can only grow in moist conditions.
NULL HYPOTHESIS: Fungi can grow anywhere and under any conditions, whether wet, dry, cold, hot, etc.

p. 23-5   Prediction—fresh, moist bread will grow more mold than dried-out bread.
Independent variable—amount of moisture
Dependent variable—growth of mold
PROCEDURE: Place several pieces of fresh and stale bread (each in its own plastic "baggie" at several locations in the house. Check the temperature of these locations. Also, all locations should have the same exposure to light. Allow one week for growth and compare samples.
OBSERVATIONS AND RESULTS: *Rhizopus* was present on the bread kept in a damp, dark environment.

Conditions for Fungal Growth Tested	Growth	No Growth
Dry bread, cool temperature		(less) √
Dry bread, warm temperature		(less) √
Damp bread, cool temperature	√	
Damp bread, warm temperature	√	

Mold or fungi require some water for growth, but can grow at various temperatures. Bread placed in the refrigerator eventually also grew mold, but not within the allotted time period. Since all bread was placed in the dark, a logical next step would be to examine the effect of light on mold growth.

Molds are commonplace due to the ability to grow or persist under diverse conditions. Though fungal hyphae require water for growth, the spores can perist for long periods of time without moisture.

Results support the hypothesis.

Results allow the null hypothesis to be rejected.

Conclusion—molds grow better when moisture is available. Molds grow better in dark areas (additional student results). Molds can grow at a variety of temperatures, although extremes of cold or heat do not support growth.

EXERCISE B   Phylum Ascomycota

p. 23-7   **a.** 8 ascospores are present in each ascus.

b. The 8 ascospores are ultimately products of mitosis. Initially, 4 ascospore nuclei are produced by meiosis. These 4 ascospore nuclei then undergo mitotsis to produce 8 ascospores.

c. *Schizosaccharomyces* asci contain a variable number of ascospores.

EXERCISE C   Phylum Basidiomycota

p. 23-9   **a.** Yes, yes, yes.

EXERCISE D   Phylum Chytridiomycota

p. 23-10   **a**   Yes.

EXERCISE E   The Deuteromycetes

p. 23-11   **a.** Mitosis.

b. *Penicillium is* the source of the antibiotic, penicillin. It is also used in making cheeses, such as blue cheese.

EXERCISE F   Identification of Collected Fungi

p. 23-13   **a.** Fungi are beneficial to humans in the decomposition of organic material, production of antibiotics, for use in making cheeses, beer and other fermented products and many (but not all) fungi are edible.

b. Fungi are harmful to humans because (1) some are saprophytes, and thus grow on human foods and cause spoilage; (2) some are plant or animal pathogens; (3) some are poisonous to humans or livestock.

EXERCISE G   Diversity Among the Lichens

p. 23-13   **a.** Most of the lichens collected come from wooded areas but they will grow on any exposed surface. However, lichens are a very sensitive indicator of toxic components of polluted air. The toxins, especially sulfur dioxide, destroy chlorophyll in the algae. Lichens absorb elements such as heavy metals and are often used to monitor industrial sites. When nuclear tests were being conducted in the atmosphere, lichens were used to monitor fallout.

b. Lichens are a mutualistic symbiosis: they are composed of both cyanobacteria or green algae and fungi so they could be classified with the cyanobacteria in the kingdom Eubacteria (except fungi do not fit in this kingdom), with green algae in the Kingdom Protista, or with the fungi in Kingdom Fungi.

c. The type of algae with which the fungus associates can be used for classifying lichens. The type of fungus is also indicative (Ascomycetes of Basidiomycetes). The physical appearance, including color, as well as habitat can be used.

PART II KINGDOM PLANTAE

NONTRACHEOPHYTES

P. 23-15 a. Mitosis
b. Meiosis. No.
c. Spores have thick coatings that can protect them from desiccation or temperature extremes. This would help to insure that genetic material is passed on to the next generation of plants.
d. Meiosis.
e. Because the embryo remains inside the mother where it is protected.
f. The gametophyte generation. Gametes are produced during the dominant generation of the life cycle in which the sporophyte and gametophyte alternate.

EXERCISE H   Nontracheophytes—Mosses, Liverworts, and Hornworts

p. 23-17 a. Sperm cells of mosses do not have flagella so they are dependent upon water currents to move the gametes (sperm) to the eggs enclosed within the archegonia.

p. 23-18 b. Mosses growing in dry environments can only reproduce during rainy weather. Spores in their protected coatings insure that the next generation of mosses will be produced.
c. Yes.
d. Eggs
e. Sperm.
f. Gametophyte
g. Haploid (*n*).
h. Spores.
i. Haploid (*n*).
i. Meiosis
k. Archegonia are cup-shaped to retain the eggs. Antheridia are feathery and shallower so that sperm can be released with rain or moisture from dew and mist.

p. 23-29 L. Archegoina are shaped to enclose and protect the egg. Antheridia are shaped to provide easy release of sperm.

1.

Division	Distinguishing characteristics	Type of hyphae	Type of reproductive structures
Zygomycota	thick-walled zygote	haploid, coenocytic hyphae	asexual meispores, sexual zygospores
Ascomycota	ascus	septate, dikaryotic hyphae	asexual conidiospores, sexual ascospores
Basidiomycota	basidium	septate, dikaryotic hyphae	basidiospores
Deuteromycota	no sexual phase	septate, haploid hyphae	asexual conidia

2. The Deuteromycetes ("fungi imperfecti") are not given phylum status because they are an artificial grouping of organisms that belong to other phyla (Ascomycota, Basiciomycota, etc.), but they share the common characteristic of lacking a sexual phase in their life cycle.

3. Chytrids are sometimes classified among the protists because they have flagellated spores and gametes. No other fungi possess flagellated cells.

4. Lichens could also be studied in association with the kingdom Protista, since most lichens are symbiotic associations of fungi with green algae. Some lichen symbionts are cyanobacteria but the eukaryotic nature of the fungal symbiont negates the possibility of studying lichens in association with the kingdom Eubacteria.

5. Lichens represent a "living together" of algae (cyanobacteria or green algae) and fungi (Basidiomycota or Ascomycota). The algae gain protection from the fungi and the fungi gain nourishment from the photosynthetic capabilities of the algae.

6. Some mosses *do* have limited amounts of unspecialized vascular tissue.
   Bryophytes—this term refers only to mosses.
   Nontracheophytes—mosses, hornworts, and liverworts. These plants do not have the *specialized* vascular tissues such as xylem (and tracheal elements) or phloem that are found in tracheophytes.
   Nonvascular plants—a term used to describe the non-tracheophytes, but if has fallen out of use because some nontracheophytes (e.g., mosses) have some unspecialized vascular tissues.

7. The green, leafy gametophyte is the conspicuous generation.

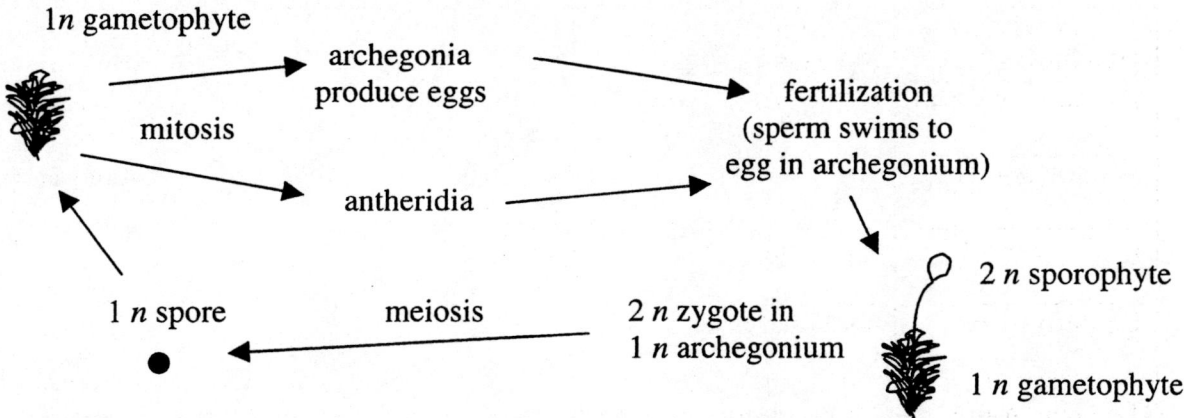

1n gametophyte

mitosis → archegonia produce eggs → fertilization (sperm swims to egg in archegonium)

antheridia →

1 n spore ● ← meiosis ← 2 n zygote in 1 n archegonium

2 n sporophyte

1 n gametophyte

8. **Rhizoids** (mosses): root-like structures which anchor the gametophyte to the soil or substrate.
   Thallus (liverworts): the gametophyte stage of a liverwort.
   Gemmae (liverworts): asexual reproduction structures in liverworts.
   Archegonium (mosses and liverworts): the female gametophyte (in which the egg develops)
   Antheridium (mosses and liverworts): the male reproductive structure (in which sperm develop).
   Spores (mosses and liverworts): haploid reproductive propagules in the asexual portion of the life cycle.

Notes:

Laboratory 24

Diversity—The Tracheophytes (Vascular Land Plants)

PART I TRACHEOPHYTES (VASCULAR PLANTS) WITHOUT SEEDS

EXERCISE A Examining Seedless Tracheophytes

p.24-2 **Table 24A-1 Vascular Plants, Divisions Psilotophyta, Lycophyta, and Sphenophyta**
Psilophyta (from top): sporangia; rhizomes; rhizoids.
Lycophyta (from top): strobili; microphylls; rhizome; roots.

p. 24-3 Spenophyta (counterclockwise from top left): microphylls; internode; node strobilus.

PHYLA PSILOPHYTA, LYCOPHYTA, AND SPENOPHTYTA

a. Lycophyta and Sphenophyta (microphyllous leaves).
b. The stems of Psilophyta can carry out photosynthesis and limited photosynthesis can occur in the stems of the Lycophyta and Sphenophyta.
c. Lycophyta have vascularized leaves.

p. 24-4 PHYLUM PTEROPHYTA
The conspicuous generation in the moss is the gametophyte generation.
a. Spores, meiosis.
b. Spore dissemination occurs in air (does not require water). Spores are actively dispersed by sporangia subjected to dry conditions.

p. 24-6 **c.** 2 cell layers
d. $2n$
e. The gametophyte degenerates.

PART II TRACHEOPHYTES (VASCULAR PLANTS) WITH SEEDS—GYMNOSPERMS AND ANGIOSPERMS

EXERCISE B Gymnosperms

p. 24-8 **Table 24B-1 Types of Conifers on Demonstration**

Conifer	Characteristics
Tsuga canadensis (eastern hemlock)	flat needles with two white bands on lower surface
Pinus elliofii (slash pine)	needles 8 -10 in. long, in bundles of 2 - 3; ovulate cone scales tipped with a spine
Pinus strobus (eastern white pine)	needles 3 - 5 in. long, in bundles of 5; ovulate cones curved, 4 - 8 in. in length
Juniperus virginiana (eastern red cedar)	branchlets covered with dark green overlapping scales; fleshy, ovulate cones
Cedrus deodora (true cedar)	needles 1 - 2 in. long, on pendant branches; ovulate cones fall apart to release seeds

p. 24-9 **a.** The male cones are much smaller. The scales are much smaller and more tightly packed.

p. 24-10 **b.** Wind.

c. Other gymnosperms are also green all year round and do not drop their leaves. Most have slender leaves that are often modified into needle-like structures. An exception to this is the ginkgo which is deciduous and has fan-shaped leaves.

EXERCISE C Angiosperms (Phylum Angiospermae)

p. 24-13 **a.** Dicot.

Figure 24C-2 *Parts of a flower*

(clockwise from top): stigma; anther; stamen; pistil; filament; sepal; ovule; ovary; style.

p. 24-14 **b.** Sucrose provides the appropriate osmotic environment for production of pollen tubes.

Laboratory Review Questions and Problems

1. The diploid sporophyte ("fronds") of a fern are the visible generation. The haploid "leafy" gametophyte of a moss is the visible generation.
2. Vascular plants have specialized conducting tissues (xylem and phloem) to transport water and nutrients throughout the plant. The presence of conducting tissues allows vascular plants to attain much larger sizes, and extend to greater heights above the water source, than non-tracheophyte plants. Additionally, the gametophyte stage is much reduced and is nutritionally dependent upon the sporophyte.
3. Yes, all contain vascular tissue. None produce seeds or fruits but rather propagete by spores. These plants are grouped as vascular plants without seeds.

4. True roots, stems, and leaves are first observed in the ferns.

5. Three major advances among the vascular plants without seeds are: (1) dominance of Sporophyte and nutritional dependence of the gametophyte; (2) no requirement for water for sexual reproduction, but rather wind or animal-mediated dispersal of sperm as pollen grains (microgametophyte); (3) a resistant seed developed from the macrogametophyte, which contains a food source for the embryo.

6. Ferns are homosporous. Gymnosperms are heterosporous. Angiosperms are heterosporous.

7.

	n or 2*n*	Gametophyte or Sporophyte
Fern archegonia	*n*	gametophyte
Moss antheridia	*n*	gametophyte
Fern leaf	2*n*	sporophyte
Moss "leaflets"	*n*	gametophyte
Pollen grain of pine	*n*	microgamethophyte
Megagametophyte of angiosperm	*n*	gametophyte
Microsporangium of angiosperm	2*n*	sporophyte
Pine tree	2*n*	sporophyte
Flower	2*n*	sporophyte

8. What types of insects or animals might serve as pollinators of flowers? How else might plants be pollinated?

   Insect pollinators will vary with the structure and color of the flowers. Insects are often used as pollinators. Bats and birds are also frequent pollinators of flowers, especially in the tropics.

   During the course of evolution, the diversification of insects greatly influenced the diversification of flower types. If a given plant species is pollinated by only one type of visitor, that flower type tends to become specialized relative to that visitor—adaptations promote constancy among pollinators. Flowers pollinated by beetles have ovaries beneath the flower petals so that the chewing of beetles will not destroy the very thing to be pollinated. Flowers pollinated by bees have bright, showy petals, usually blue or yellow. They are often patterned—even with distinct markings, "honey guides" that the human eye cannot see. Some of the flowers possess complex traps or passageways that force bees, as they feed on nectar, to brush against the pollen-laden anthers. Moths, which are nocturnal, feed on flowers that are usually white with heavy fragrances. Usually the nectaries of moth- or butterfly-pollinated flowers are located at the end of a long corolla tube that requires the pollinator to have long, sucking mouth parts. Bird-pollinated flowers are usually red or yellow and are part of large inflorescenses.

   Grass flowers are wind pollinated. Flowers that are wind pollinated are usually small and inconspicuous or uncolored and not patterned.

   The evolution of plant and pollinator adaptations provide graphic examples of coevolution.

9. You may wish to add this question if students completed the optional work on Figure 24B-4 and Figure 24C-5. Answers are included in the table below.

Compare angiosperms and gymnosperms by identifying each structure by it common name and location in relation to other plant structures.

General Term	Description	
	Gymnosperms	Angiosperms
Sporophyte	tree	flower
Microsporophyll	scale of male cone	stamens evolved from structure
Microsporangium	attached to scale—contain developing microspores	pollen sacs
Microspores	developing pollen grains	developing pollen grains
Megasporophyll	scale of female cone	carpels evolved from sstructure
Megasporangium	contained within ovules	within ovules of ovary
Megaspores	forms multicellular megagametophyte	gives rise to multinucleate megagametophyte
Gametophyte	microgametophyte—mature pollen megagametophyte—contains archegoina with eggs	microgametophyte—two-celled pollen grain Megagametophyte—embryo sac with 8 nuclei

Laboratory 25

Diversity—Porifera, Cnidaria, and Wormlike Invertebrates

p. 25-3   **a.** Calcareous sponges (calcium) and glass sponges (silicon)
           b. Most of the soft specimens on display have fibrous skeletons.
           c. Water outflow

p. 25-4   **d.** Filter feeding
           e. Cytoplasmic collars and flagellae extend from the choanocytes.
           f. Food particles become trapped on the cytoplasmic collar of the choanocytes and the flagellum assists in directing the water flow through the sponge body.
           g. The region is jelly-like. It contains skeletal elements of SiO_2 or $CaCO_3$ or spongin protein.
           h. Amoebocytes.
           i. Small, needle-like structures.
           j. Spicules.
           k. Vinegar (an acid) or bleach (a base) dissolve the cells of the sponge body so you can observe the spicules.

EXERCISE B   **Phyla Cnidaria and Ctenophora**

p. 25-7   4. Nematocysts assist in attaching the *Hydra*'s tentacles to the substrate as it moves. They are also used to trap prey by piercing, sticking to the prey, or wrapping around the prey.

           HYPOTHESIS: If *Hydra* are presented with both living and dead brine shrimp, they will eat only the live food.

           NULL HYPOTHESIS: Whether prey are alive or dead makes no difference in their use by *Hydra* as a food source.

           Prediction—*Hydra* will respond to live, moving food, but not to dead food.

           Independent Variable—live or dead brine shrimp.

           Dependent Variable—number eaten (used as food).

           PROCEDURE:

               Place 10 *Hydra* in a small Petri dish. Add 5 drops of brine shrimp culture. Observe for 2 minutes. Record the number of captured prey/feeding *Hydra*. Repeat, using 5 drops of heated (and cooled) brine shrimp culture (brine shrimp are dead).

           RESULTS: All 10 *Hydra* attacked and began the process of eating live *Hydra*. In the dish containing dead prey, the *Hydra* caught the prey when they bumped into the tentacles, but did not eat them.

           Results support the hypothesis but reject the null hypothesis.

           The prediction is correct.

Conclusion—*Hydra* prefer live food. Note: while touch causes cnidocytes to fire nematocysts, amino acids and other chemicals such as glutathione must be released by prey to cause tentacles to contract and the *Hydra* to eat.

p. 25-9   **a.** Medusa buds produce the free-swimming medusae which produce sperm or eggs. Thus, medusa buds initiate the sexual reproduction phase of the life cycle and disperse to new areas.

b. The colonial life form allows for cooperation between specialized types of polyps (feeding, reproductive).

p. 25-10   **c.** Both have two-layered (polyp-type) bodies with a basal disk and tentacles.

d. Soft corals like sea fans and sea whips are similar to other corals in that the polyps live within tiny holes in the coral skeleton.

e. Organisms in the phylum Ctenophora are distinguished from organisms in the phylum Cnidaria by their eight longitudinal, "comb-like" bands of fused cilia, and by the lack of stinging cells in ctenophores.

EXERCISE C   Wormlike Animals: Platyhelminthes, Rhyncocoela, Nematoda, Annelida

PART 1   Phylum Platyhelminthes: Flatworms

p. 25-11   **a.** Bilateral symmetry.

b. Pharynx; the egg yolk (dyed red) begins to spread out to fill all branches of the digestive (gastrovascular) cavity.

c. Through the pharynx.

p. 25-13   **5.** HYPOTHESIS: If the eyespots of *Dugesia* are used to sense light, they will demonstrate a positive phototaxis. (Note: this is a typical student hypothesis—results will not support this hypothesis.)

NULL HYPOTHESIS: *Dugesia* show no preference for light or dark.

Prediction—*Dugesia,* given the choice between light and dark, choose light.

Independent Variable—the amount of illumination.

Dependent Variable—*Dugesia*'s direction of movement.

RESULTS: Planaria move to the side of the Petri dish covered by aluminum foil and do not reappear during the course of the experiment.

Light   Dark

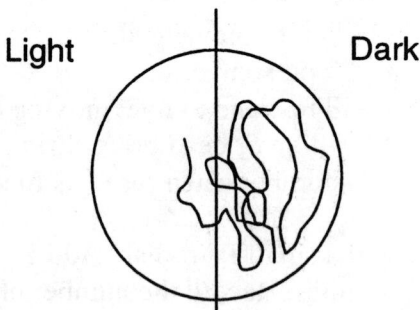

Conclusions—Eyespots (or other sensors) respond to light and make it possible for *Dugesia* to avoid light (show negative phototaxis). The original hypothesis is not supported. The null hypothesis is also falsified because planaria move to the dark. A new hypothesis should be formulated and tested.

The original prediction was not correct.

p. 25-14   **d.** Loss of sense organs is a secondary effect of the endoparasitic mode of life. They

live in a very constant environment inside their host.
  e. Snails are the intermediate host. Fish and humans are the definitive hosts.
  f. Hooks and suckers on the scolex attach it to the host intestine.

p. 25-15 g. The thick cuticle protects the tapeworm from digestive enzymes secreted within the host intestine.
  h. No, all proglottids do not have the same structure, and some are larger than others. The proglottids at the posterior end of the tapeworm were formed first.
  i. Both male (testes) and female (ovary, uterus)-reproductive structures are apparent.
  j. A hermaphoditic species has the advantage of continuos and efficient reproduction. A mate is always available and no energy is wasted in searching for a mate. New variations introduced by recombination are also reduced. Many hermaphrodites live in relatively unchanging environments where variation may lead to less well adapted offspring so reducing variation may be advantageous.

PART 2 Phylum Rhynchocoela: Ribbon Worms

p. 25-15 Preserved ribbon worms are long and flattened. The extended proboscis is barbed. External segments are not evident.

PART 3 Phylum Nematoda (Roundworms) and Other Wormlike Phyla

p. 25-18 a. Adult *Trichinella* are minute roundworms. *Trichinella* larvae increase to a size ten times that of adults and encyst, becoming encased in a thick wall formed by the host tissue.
  b. Encysted *Trichinella* are protected by the tough coat of the cyst. High heat is necessary to kill them.
  c. Small children may eat dirt or inadvertently introduce dirt into their mouths from an area infected with *Ascaris* eggs (e.g., a barn yard where pig feces could have been tracked through on shoes). Handling the worms themselves might also transfer eggs.

p. 25-20 d. Lymph capillaries penetrate many of our tissues (they are absent in the central nervous system and bones and teeth). They collect filtrate from the circulatory system and tissue fluids and return them to the heart via the circulatory system. Lymph vessels often parallel arteries and veins.
  e. Infection of the guinea worm can be controlled by treating drinking water to remove *Cyclops,* a crustacean that serves as the intermediate host. Failing this, people should be warned that individuals with open blisters formed by the adult worms should not bathe or wash clothes in water also used for drinking.

Extending Your Investigation: Nematode Diversity

HYPOTHESIS: A greater number of nematodes is present in moist, organic soil than in dry, sandy, or clay soils.

NULL HYPOTHESIS: Nematodes will be evenly distributed in different types of soil.

Prediction—More nematodes will be extracted from soil out of the vegetable garden or woods than from clay soil taken from between grass patches on the lawn.

Independent Variable—soil type.

Dependent Variable—number of nematodes in the sample.

p. 25-21 RESULTS:

Soil Type	Number of Nematodes/ml
botanical garden (sandy)	1
lawn (clay)	0
vegetable garden	2
woods—near rotting log	5

Results support the hypothesis but the null hypothesis is rejected.
The prediction was correct.
Conclusion—nematodes are found in rich organic soils.

PART 4 Phylum Annelida: Segmented Worms

p. 25-22 **a.** Yes, but the thickening of the body wall in the region of the clitellum obscures the segmental grooves.

b. Laboratory specimens had 60-90 segments. All earthworms do not have the same number of segments.

p. 25-19 **Figure 25C-10** *Label the structures visible in the earthworm cross section*
a. intestine; **b.** typhosole; **c.** muscle tissue; **d.** chlorogen tissue; **e.** coelom; **f.** nerve cord; **g.** blood vessels; **h.** nephridium; **i.** longitudinal muscles; **j.** circular muscles; **k.** epidermis; **1.** cuticle; **m.** seta.

p. 25-25 **c.** Coelomic fluid.

d. The coelom serves as a hydrostatic skeleton.

Laboratory Review Questions and Problems

1. Metazoa: multicellular motile, heterotrophic organisms that exhibit the blastula stage during development

Parazoa: loose aggregations of specialized cells; because of their isolated phylogenetic position, they are often described as distinct from other metazoans that are then referred to as "eumetazoa"

Eumetazoa: "true" Metazoa; distinct from the sponges (Parazoa)

Protozoa: single-celled, animal-like protists

2.

	Hydrozoans	Scyphozoans	Anthozoans	Ctenophorans
Polyp or medusa as predominant form	polyps, some medusae	predominantly medusae	polyps only	
Presence or absence of cnidocytes	present	present	present	absent
Distinguishing characteristics	usually small	large medusa, may be brightly colored	partitioned gastrovascular cavity	8 longitudinal rows of ciliated "combs"
Examples	*Obelia, Hydra, Physalia*	*Aurelia*	*Metridium*, sea anemones	*Pleurobranchia*

3. Cnidocytes contain stinging organelles called nematocysts. All of the organisms in the Phylum Cnidaria have these stinging cells.

4. Biradial symmetry is characteristic of the ctenophores. All but one species lack nematocysts.

5. Platyhelminthes have bilateral body symmetry and an acoelomate body cavity.

6. Ribbon worms' protonephridia are truly excretory (removing nitrogen-containing compounds) in function, whereas flatworms' protonephridia are used only for osmoregulation. The structure of the digestive tract differs in that ribbon worms have two digestive openings (mouth and anus) whereas flatworms have a single opening, the pharynx, for intake and excretion.

7. Annelida, schizocoelous; Nematoda, pseudocoelous; Platyhelminthes, acoelous; Rhynchocoela, pseudocoelous ("rhynchocoel"); Chordata, enterocoelous. The line should be drawn below the echinoderms, chordates, and annelids, just below the split in the trunk of the diagram.

8. Flatworms do not have a coelomic cavity, rather the body is filled with a continuous mass of mesoderm and these organisms are considered to be acoelous. Roundworms have a body cavity only partially lined by mesoderm and are called pseudocoelous. Rhynchocoels possess a body cavity often called a rhynchocoel, but it is lined by endoderm so does not fit the definition of a true coelom—a body cavity lined by mesoderm. Segmented worms are true coelomates and the mesodermal liming of the body cavity is formed by a "splitting" of the mesoderm (hence, schizocoelous) rather than from an outpocketing (evagination) of the developing gut (enteron) as is true of echinoderms and chordates (enterocoelous coelomates).

9. Porifera have only two layers of body tissues, epidermis and endodermis. Cnidaria and Ctenophora have two layers plus a mesoglea that is a jelly-like layer containing some loose mesenchyme (diploblastic). All other organisms develop from three body tissue layers: ectoderm, mesoderm, and endoderm and are known as triploblastic.

10. Porifera are asymmetric. Cnidaria and Ctenophora are radially symmetric. All other organisms show bilateral symmetry. (Echinoderms are secondarily radial; echinoderm larvae exhibit bilateral symmetry.)

	Porifera	Cnidaria	Platyhelminthes	Nematoda	Annelida	Evolutionary Trend
Number of tissue layers	2: ectoderm; endoderm	2: ectoderm; endoderm; acellular mesoglea	3: ectoderm; endoderm mesoderm	3: ectoderm endoderm mesoderm	3: ectoderm; endoderm; mesoderm	3 tissue layers
Type of digestive cavity	gastrovascular	gastrovascular (one opening for mouth/anus)	gastrovascular (one opening for mouth/anus)	"true gut" (mouth and anus)	complete gut (one-way digestive tract)	one-way digestive tract with specialized regions for digestion and absorption
Type of coelom (if present)	acoelomate	acoelomate	acoelomate	psuedocoelomate (hydrostatic skeleton)	segmented coelom (hydrostatic skeleton)	partitioned coelom
Type of reproduction (asexual /sexual)	asexual: fragmentation; gemmules; buds. sexual: monoecious	asexual: buds sexual: monoecious or dioecious	asexual: regeneration; also by larval stages of parasites sexual: monoecious	sexual: usually dioecious	sexual: monoecious	separate sexes (dioecious); sexual reproduction
Larva (if discussed)	2 types	planula			trochophore	various adaptive solutions
Nervous system (form and location)	none	nerve net; longitudinal nerve cords	first central nervous system: anterior ganglion running posteriorly	nerve ring at anterior end with two nerve cords	centralized nervous system: paired ganglia in each segment	cephalization and a central nervous system
Sensors	no sense cells; ocelli	sensory cells; statocysts; ocelli; touch	sensory cells; statoreceptors, ocelli; touch	sensory cells; chemoreceptors	sensory cells; ocelli	increasingly complex sensory organs

11 (cont.).

	Porifera	Cnidaria	Platyhelminthes	Nematoda	Annelida	Evolutionary Trend
Type of circulation (open or closed)	none	none	none	none	closed	closed circulation, especially as size increases, to control composition of body fluids
Type of excretory organs (structure; osmoretulatory or excretory functions)	individual cells	individual cells	protonephridia (flame cells)	protonephridia	paired metanephridia (some have protonephridia)	metanephridia, enhanced ultrafiltration
Type of symmetry	asymmetry or radial	radial	bilateral	bilateral	bilateral	bilateral

Notes:

Laboratory 26

Diversity—Mollusks, Arthropods, and Echinoderms

PART 1   Chitons

p. 26-3   **a.** Eight plates.

PART 2   Bivalves

p. 26-3   **a.** A clam is a bivalve.

b. The smallest of the concentric rings, on the part of the shell nearest the umbo (beak) form first.

p. 26-4   **Figure 26A-2(b) Internal structure of the clam**

a. mantle; **b.** incurrent siphon; **c.** excurrent siphon; **d.** gills; **e.** attachment site for adductor and retractor muscles **f.** mouth; **g.** palps; **h.** stomach; **i.** digestive gland; **j.** intestine; **k.** excretory organ; **1.** heart; **m.** anus.

p. 26-5   **c.** First, the adductor muscles contract. Water expelled from the mantle cavity loosens the sand. Water and blood act as a hydrostatic skeleton, dilating the foot and anchoring it in the sand. Finally, the anterior and posterior muscles contract, pulling the mollusk down into the sand.

d. Other representatives include: oysters; scallops; other clams.

Extending Your Investigation: Filter Feeding

p. 26-6   HYPOTHESIS: The clam uses its gills to trap microscopic food in a mucous net. This net is then moved to the mouth and ingested.

NULL HYPOTHESIS: There is no sequential involvement of specific parts or pattern involved in feeding.

Prediction—small particles will be trapped by the mucous net on the gills and ingested.

Independent variable—food particles.

Dependent variable—pathway taken

Procedure—follow the particles on the gill surface using a dissecting microscope.

RESULTS: Particles are trapped in a net of thick mucus secreted by the gills and moved toward the ventral edges. They are then passed anteriorly where ingestion may be assisted by the palps.

Results support the hypothesis; results lead to rejection of the null hypothesis.

The prediction was correct.

Conclusions—clams are filter feeders. They use their gills and a mucous net to trap food particles that are then passed into the digestive tract.

PART 3   Gastropods

p. 26-6   **a.** To the right

p. 26-7   **b.** The head and foot

c. Yes. The head is well-developed, with eyes and sensory tentacles.

d. Yes, the operculum completely seals off the body. No, the operculum is not continuous with the shell.

e. When the snail is crawling, the operculum is on the dorsal side of the foot.

PART 4   Cephalopods

p. 26-8   **a.** The arms.

PART 5   The Scaphopoda

EXERCISE B   Phylum Arthropoda

PART 1   Crustacea: External Morphology of the Crayfish

p. 26-11   **Table 26B-1 Comparison of the External Morphology of Arthropods**

Characteristics	Crustaceans	Arachnids	Insects	Myriapods
Main divisions of the body	cephalothorax, abdomen	cephalothorax, abdomen	head, thorax, abdomen	head, thorax, abdomen
Main body divisions that show external segmentation	abdomen	none in spiders, abdomen in scorpions	thorax, abdomen	abdomen
Locomotor structures	5 pairs of appendages for crawling/ digging; 6 pairs for swimming	8 legs (4 pairs)	6 legs (3 pairs)	2 legs on each of the many segments
Other appendages		1 pair of pedipalps, 4 pairs of legs	wings	
Number of pairs of antennae	2 pairs	no antennae	1 pair	1 pair
Mouthparts	mandibles, maxillae, maxillipeds	chelicerae (distal portion modified to form fangs or pincers; labium	mandibles. maxillae, labium, labrum	epistome and labium; maxillae; mandibles; hypopharynx
Number and type of eyes	2 compound eyes on stalks	8 simple eyes	2 compound eyes and ocelli	simple eyes (ocelli)

PART 2   Arachnida: External Morphology of a Spider

p. 26-12   **a.** The cephalothorax includes the head, has legs attached, and is smaller than the abdomen.

b. Cephalothorax

c. 8 eyes (fewer in some species). These eyes are arranged in two rows of 4 along

the anterior dorsal margin of the carapace.
  d. The anterior lateral (black) pair of eyes is the largest.

p. 26-13 **e.** Most spiders have six spinnerets (one set is vestigial or modified) on the posterior of the abdomen. Primitive spiders have eight spinnerets in a more anterior position.

PART 3   Chelicerata: Biochemical Taxonomy of the Horseshoe Crab (Optional)

p. 26-14   **Table 26B-2**

Note: (1) the entire lactic acid solution must be adjusted to pH 7.0, not just the 0.01 M NAD, if this exercise is to work correctly. see page 26-3 **(2)** of the *Preparator's Guide;* (2) absorbance values and rapidity of decolonization are *very* dependent upon the potency of the extracts)

	Spider		*Limulus*		Crab		Control	
	Abs.	**Color**	**Abs.**	**Color**	**Abs.**	**Color**	**Abs.**	**Color**
Assay mix	0	blue	0	blue	0	blue	0	blue
Assay mix + indicator	0.289	blue	0.194	blue	0.297	blue	0.321	blue
Assay mix + indicator + homogenate initial reading	0.295	blue	0.285	blue	0.291	blue	0.299	blue
5 minutes	0.233	medium blue	0.195	medium blue	0.123	pale blue	0.294	blue
10 minutes	0.175	pale blue	0.111	pale blue	0.015	very pale blue	0.288	blue
15 minutes	0.149	very pale blue	0.025	very pale blue	−0.009	clear	0.282	blue

p. 26-15   **a.** *Limulus is* more closely related to the spiders, Class Arachnida, because the spider and *Limulus* extracts decolorized the indicator mix more slowly than did crab extract.

  b. Horseshoe crabs are superficially anatomically similar to crabs, but differ in that true crabs have eight pairs of walking legs, whereas the horseshoe crab has only 5 pairs.

  c. Both *Limulus* and spiders belong to the subphylum Chelicerata. Ancestors of both were aquatic. *Limulus* has remained within the marine environment. Living spiders evolved to become almost exclusively terrestrial. Since horseshoe crabs and true crabs are adapted to the marine environment, both display anatomical modifications for this habitat.

PART 4   Insecta: External Morphology of the Grasshopper

p. 26-15   **Figure 26B-4 External anatomy of the grasshopper**
    a head; **b.** thorax; **c.** abdomen; **d.** antennae; **e.** labrum; **f.** mandibles; **g.** labium; **h.** sensory palps; **i.** femur; **j.** typanum (sense organ)

p. 26-16   **a.** The insect body is divided into three basic parts, head, thorax and abdomen, whereas crustacean and arachnid bodies are divided into two parts, cephalothorax and abdomen.

    b. Three ocelli are located between the eyes and the antennae.

    c. One pair of spiracles (one spiracle on each lateral surface) is present on each segment.

PART 5   Diplopoda and Chilopoda

PART 6   Insect Mobility

p. 26-17   **a.** The insect having the greatest mass is the "goliath beetle" from Africa. The insect with the largest wingspan is the "atlas moth" from southeast Asia.

    b. Spiracles are located on the lateral surfaces (sides) of the first 7 or 8 segments.

    c. Chitin is arranged in spirals within the spiracle.

    d. The hairs serve as filters, preventing entrance of dust or parasites into the tracheal system.

    e. Chitin is arranged in "spirals" (helices) within the individual trachae and tracheoles.

    f. The helical arrangement strengthens the trachae and tracheoles.

    g. The tracheal system has multiple branches.

  3. **Table 26B-3**

No Wings	One Pair of Wings	Two Pairs of Wings

silverfish
earwig (may have one
  pair)
flea
louse
	beetle (second pair modified as
	  protective plates, elytra)
	flies, gnats, mosquitoes (second
	  pair reduced to knobs, halteres,
	  which aid in stabilizing flight)
		dragonfly
		butterfly
		bee, wasp, and ant
		grasshopper

    h. Wings extend from the thorax

    i. The wings cover the abdomen.

    j. The wings serve a protective secondary function.

EXERCISE C   Protostomes: Minor Groups

EXERCISE D   Phylum Echinodermata

p. 26-19   **a.** All adult forms of echinoderms are radially symmetrical but this is secondary to their original bilateral larval origins. All echinoderms are bilateral organisms.

    b. The dipleurula larvae are bilaterally symmetrical.

    c. Starfishes (class Stelleroidea) have arms

    d. Sea urchins (class Echinodea) are spherical.

    e. Sand dollars (class Echinodea) are disclike.

p. 26-20   **f.** Sand dollars have 5 teeth; this is consistent with their five-part symmetry.

1.

	Protosomes	**Deuterostomes**
Type of coelom	schizocoelous	enterocoelous
Type of cleavage	spiral	radial
Type of development	determinate	indeterminate
Fate of the blastopore	mouth	anus

2. Chitons are mollusks. Most mollusks do not show external segmentation. Organisms in the annelid and arthropod phyla are externally segmented. Thus the chitons, with a basic molluscan body plan, yet displaying external segmentation, are grouped with annelids and arthropods.

3. The anterior region of the foot is modified to form the "arms" which are adapted for capturing prey. The mantle cavity is modified for swimming by propulsion. Water taken into the mantle cavity is forcibly expelled through a ventral tubular funnel, thus propelling the cephalopod in the opposite direction.

4. The arthropods are unified by the characteristics of: a chitonous exoskeleton, divided into plates connected by articular membranes (joints); periodic shedding (molting) of the exoskeleton and growth through a succession of juvenile forms (metamorphosis); the blood vascular system (hemocoel) is open, the coelom is greatly reduced.

5. Sample dichotomous key (other forms are possible):
   la. Body in two parts (cephalothorax and abdomen).. 2
   lb. Body in three parts (head, thorax, abdomen) 3
      2a. One pair of antennae ....Arachnida
      2b. Two pairs of antennae .... Crustacea
         3a. Many legs 4
         3b. Six legs (three pairs)... Insecta
            4a. Two legs per segment...Chilopoda
            4b. Four legs (two pairs) per segment .... Diplopoda

6. Crustaceans are distinguished from other arthropods by: two pairs of antennae and biramous appendages.

7. Crustaceans and Arachnids have a two-part body (cephalothorax and abdomen). Arthropods have a three-part body (head, thorax and abdomen).

8. Echinoderns differ in: radial cleavage; enterocoelous development; internal skeletons; pentamerous radial symmetry (secondarily derived); coelomic canals (water-vascular system); large coelom; well-developed digestive tract; and no excretory system.

9. The four classes of echinoderms share the characteristics of pentamerous (five-part) radial symmetry, spiny protective skins, and a water vascular system (tube feet).

6

Notes:

Laboratory 27

Diversity—Phylum Chordata

EXERCISE A   Hemichordates

p. 27-5   **a.** The presence of a stomochord and a body consisting of three sections: proboscis; collar; and trunk, each with its own coelomic cavity.

         b. Pharyngeal slits.

EXERCISE B   Tunicates

p. 27-7   **a.** Tunicate larvae have a notochord, a dorsal tubular nerve chord, pharyngeal pouches, and a post-anal tail. Adults retain only the pharyngeal pouches (as slits that perforate the pharyngeal wall and filter water passing from the pharynx to the atrium).

EXERCISE C   Sea Lancelets

p. 27-7   **a.** Bilateral (the organism can be split longitudinally into two mirror-image halves).

p. 27-8   **b.** The notochord is alternately flexed by the muscles on opposite sides to produce undulating swimming movements. Without this incompressible element, the muscles could not propel the organism forward.

p. 27-10   **c.** Cephalochordates are sessile filter-feeders and therefore do not need sense organs to obtain food. It is not clear whether this condition is primitive or derived but, considering that tunicate larvae have a cerebral vesicle with prominent sensors, it is logical to assume that the lack of an expanded anterior "brain" with sensors is somehow a specialization.

EXERCISE D   Vertebrates—Fishes

p. 27-10   **a.** Dogfishes and teleosts have two pairs of paired fins.

         b. Dogfishes' fins are fleshy. In more advanced teleost fishes like the perch, the pelvic fins may move anteriorly. Sharks have two dorsal fins. Bony fishes have 1-2 (or more) dorsal fins. Some bony fishes like the perch also have unpaired anal fins.

p. 27-11   **Table 27D-1 Comparison of the External Morphology of Three Fishes**

	Lamprey	Dogfish	Teleost
Features of the integument	leathery (ancient ostracaderms were armored)	dermal denticles (no bone) and spines	dermal scales (reduced bone)
Number of fins (paired or unpaired)	2 fins (and 2 ventral metapleural folds)	2 paired, 3 unpaired	2 paired, 2-4 unpaired

Table 27D-1 (cont.)

	Lamprey	Dogfish	Teleost
Location of fins	unpaired caudal and dorsal fins	pectoral and pelvic pairs; 2 unpaired dorsal and caudal fins	pectoral and pelvic pairs; 1-2 unpaired dorsal fins, and anal and caudal fins
Number of external nares	1	2	2
Position of external nares	median, dorsal, anterior	anterior, dorsal	anterior, above jaw
Texture, size and shape of the teeth	"horny" teeth (cornified epithelium)	dentine	dentine
Position of the teeth inside the mouth	one set on the walls of the mouth; one set on the tongue	around mouth in jaws	in jaws (and palate, pharynx, etc.)
Number of gill slits	7 (from 5 to 15 in hagfishes)	6 (including spiracle)	5
Location of gill slits (covered or uncovered)	external (uncovered)	external (uncovered)	internal (covered by an operculum)
Other distinguishing features observed	no jaws; cartilage skeleton	jaws; cartilaginous skeleton	jaws; air bladder; bony skeleton

    c. Lampreys are highly specialized ectoparasites. Dogfishes are carnivorous. However, the number and position of the nares are more likely related to the diverse ancestry (monorhine vs. diplorhine) and developmental state (relative amount of tissue incorporated into the mouth) of these very different fishes. The nares do not open into the pharynx in either the lamprey or shark.

EXERCISE E   Amphibia and Reptiles

p. 27-12 **a.** Amphibian skins are wet and may have a mucous coat making them slippery. Reptile skins are dry and covered by epidermal scales.

    b. The keratinized scales of the reptilian skin offer protection from water loss and drying (desiccation).

    c. Oval.

    d. Scales are smaller and more numerous in the tail and feet.

    e. This arrangement affords maximum flexibility for movement.

    f. Amphibian feet are webbed. Reptiles have claws.

    g. The webbed feet betray the aquatic heritage.

    h. Amphibian limbs are oriented lateral to the body. Reptile limbs are directed more ventrally.

    i. Laterally-oriented limbs are useful for swimming, ventrally-oriented limbs for walking/running.

p. 27-13 **j.** The metacarpals are more compact in turtles, and the radius and ulna are fused.

Frog limbs are modified for jumping and swimming.

k. The angle in the turtle is about 90°. The angle in the frog is acute.

1. The metatarsals in the turtle are reduced and more compact. In the frog, the tibia and fibula are fused.

m. The hindlimbs differ in length and sturdiness (compactness).

n. Frog limbs are adapted for jumping, turtle limbs for crawling.

p. 27-14 **o.** Reptilian eggs are leathery and larger. Amphibian eggs are gelatinous and small.

p. Gelatin, a "jelly" coat.

q. In frogs, fertilization is external (the male grasps the female tightly—amplexus—and releases sperm as she lays eggs. In reptiles, fertilization is internal (a penis in turtles, hemipenes in snakes and lizards). Amphibian eggs are aquatic (or must be kept moist during development), reptilian eggs are terrestrial.

EXERCISE F   The Avian and Mammalian Skeletal Systems

p. 27-16 **a.** The pigeon skull appears to be relatively larger (to accommodate very large eyes and a well-developed brain).

b. Bird bones are less dense ("lighter"). They contain air chambers (which connect to the pulmonary system of air sacs) crossed by supporting bony trabeculae.

c. The "spongy" composition of the bones.

d. Cervical, lumbar, sacral, caudal (coccyx)

e. *Cat* (52±): 7 cervical; 13 thoracic; 7 lumbar; 3 sacral (fused to form sacrum); about 22 caudal. *Humans* (33±): 7 cervical; 12 thoracic; 5 lumbar; 5 sacral (fused to form the sacrum); 4 (3-5) coccygeal (fused to form the coccyx). *Pigeon* (42±): 14 cervical (11-25 in other birds); 5 thoracic (3 fused together, 5th fused with synsacrum); 8 lumbar and sacral (fused to form syncacrum); 15 caudal (last 5 form terminal pygostyle which supports the tail feathers).

f. The greater number of cervical vertebrae and their saddle-like articulation allows for a longer, more flexible neck, for feeding (most of the remainder of the axial skeleton in birds is tightly fused).

g. Clavicles + interclavicle = furculum or wishbone.

h. Shoulders. They brace the scapula (and forelimb) against the sternum (which, in turn, is connected to the vertebral column by the ribs).

i. Ischium, pubis, illium. The same bones are found in the pigeons. In both mammals and birds, the pelvic bones fuse to each other and to the vertebral column to form the sacrum. The acetabulum is found at the junction of the three and accommodates the head of the femur. Note that these bones fuse ventrally (pubic symphysis) in mammals but remain open in egg-laying birds.

j. In birds, the femur is lateral to the pelvis. In mammals, the femur is beneath the pelvis.

k. Weight distribution and transference of force from the hindlimb to the axial skeleton.

p. 27-17 **1.** The more rigid structures provided by fused bones are used for support for muscle contraction during flight, restrict free movements of the elements, and weigh less than the individual bones.

Table (see below—Laboratory Review Questions and Problems, question # 2.

p. 27-18   **m.** Hindlimbs of birds are used primarily for perching and hopping, not for running, as in many mammals. A further difference, related to maintaining the center of gravity above the feet in birds when they roost or perch, is the relocation of the "ankle" joint from the tibio-tarsus to an "intertarsal" (mesotarsal) joint found in birds (note the ending of the "splint-bone" or fibula alongside the tibia in birds—the portion of the "tibia" beyond this point consists of fused tarsals and the ankle joint is formed between the proximal tibio-tarsus and distal tarsals of the foot).

Laboratory Review Questions and Problems

1. Homologous structures have the same embryological or evolutionary origin and a related genetic base. Examples: (1) bones of forelimb of turtle and frog and wing of bird; (2) bones of hindlimb of turtle and frog; (3) spiracle of dogfish and outer / middle ear and eutachian tube of tetrapods.

2. (on previous page, p. 27-27)

Class	Chordate characteristics present	Chordate characteristics missing	Invertebrate or vertebrate	Distinguishing characteristics	Adaptations to environment
Urochordata	all present in larva (notochord, dorsal tubular nerve cord, pharyngeal slits, postanal tail,)	all (except pharyngeal slits) missing in adult	invertebrate	tunicin test; unusual blood pigments; reduced coelom; hermaphroditic	filter feeder; sucker on head for attachment; reflex contractions when disturbed
Cephalochordata	all	none	invertebrate	nerve cord extends to the front of the head	wheel organ guards the pharynx; metaplueral folds act as "fins"
Agnatha	all	none	vertebrate	no jaws; single nostril; cartilaginous skeleton	lamprey is an ectoparasite (grasping mouth, rasping teeth)
Chondrichthyes	all	none	vertebrate	cartilaginous skeleton, dermal scales; gill slits open separately	cartilaginous skeleton reduces density; teeth for carnivory
Osteichthyes	all	none	vertebrate	bony skeleton, dermal scales; gills covered by operculum in teleosts	lung/swim bladder for respiration and buoyancy
Amphibia	all	none	vertebrate	moist skin used for respiration; poison glands	aquatic reproduction ("tadpoles")

2. Cont.

Class	Chordate characteristics present	Chordate characteristics missing	Invertebrate or vertebrate	Distinguishing characteristics	Adaptations to environment
Reptilia	all	none	vertebrate	epidermal scales	land egg, impermeable skin (keratin)
Aves	all	none	vertebrate	endothermy, feathers; flight	land egg, skeletal adaptations for flight
Mammalia	all	none	vertebrate	endothermy, hair; mammary glands	adapted to diverse environments

Notes:

Laboratory 28

Plant Anatomy—
Roots, Stems, and Leaves

EXERCISE A	**Plant Tissues**	

p. 28-4   **a.** Vascular tissue (xylem)
           b. Ground tissue (sclerenchyma)

EXERCISE B	**The Monocot and Dicot Angiosperm Body Plan**

PART 1   Structure of the Dicot and Monocot Root

A   The Herbaceous Dicot Root—Buttercup (*Ranunculus*)

p. 28-7   **Figure 28B-2** *Cross section of a dicot root*
           (a) epidermis; (b) cortex); (c) cortical parenchyma; (d) endodermis; (e) stele; (f) pericycle; (g) primary xylem; (h) primary phloem; (i) vascular cambium

B   The Monocot Root—Corn (*Zea mays*)

p. 28-8   **Figure 28B-3** *Cross section of a monocot root*
           (a) epidermis; (b) cortex; (c) stele; (d) pith
           a. Pericycle
           b. Facilitates water transport through the xylem

PART 2   Structure of the Dicot and Monocot Stem

A   The Herbaceous Dicot Stem—Alfalfa (*Medicago sativa*)

p. 28-9   **Figure 28B-4** *Cross section of a dicot stem*
           (a) epidermis; (b) cortex; (c) parenchyma; (e) collenchyma; (f) pith; (g) vascular bundles; (h) phloem; (i) xylem; (j^1) vascular cambium; (j^2) vascular cambium

B   The Monocot Stem—Corn

p. 28-10   **Figure 28B-5** *Cross section of a monocot stem*
           (a) epidermis; (b) sclerenchyma; (c) parenchyma; (d) vascular bundles; (e) phloem sieve tubes; (f) xylem; (g) phloem sclerenchyma

C   The *Coleus* Root and Stem

p. 28-11   **a** Dicot
           b. The distinguishing root characteristics are the lack of pith and the presence of a solid stele, composed of the vascular tissues and pericycle surrounded by the endodermis. Monocot roots have pith, with strands of primary xylem and phloem arranged in a cylinder outside the pith.

p. 28-12   **c.** The distinguishing stem characteristic is the arrangement of vascular bundles (each composed of xylem and phloem) in a circle around the central pith. Monocot stems have vascular bundles scattered throughout the stem cross section.

d. No. Parenchyma and collenchyma cells stain-red-purple. Phloem stains red. Xylem is bluish-green or green. Sclerenchyma tends to be blue-green

Extending Your Investigation: Is It a Monocot or Dicot?

p. 28-12 HYPOTHESIS: If the arrangement of vascular bundles in the stems of houseplants and weeds can be examined, they will exhibit the structure of typical dicots.

NULL HYPOTHESIS: There will be no difference in the arrangement of vascular bundles in the stems of plants no matter what type.

Prediction—vascular bundles in the materials studied will appear to be arranged around the periphery of the stem—typical of dicots.

Independent variable—plant type.

Dependent variable—arrangement of vascular bundles.

PROCEDURE: Use the Toluidine Blue O staining procedure outlined in Exercise B, Part 2C.

p. 28-13 RESULTS: Begonias, geraniums, *Zebrina,* clover, and sunflowers were all dicots. Onion grass was a monocot.

Observations support the hypothesis and allow us to reject the null hypothesis.

The prediction was correct.

Conclusion—most house plants are dicots. Plants in the grass family are monocots.

The sample size is too small to make this a strong or general conclusion for all house plants.

PART 3   Leaf Structure

A   External Features of Leaves

B   The Structure of a Dicot Leaf—Privet (*Lingustrum*)

p. 28-14 **a.** The cuticle retards water loss.

b. Xylem is oriented towards the upper surface; phloem is oriented towards the lower surface. In the stem, xylem is on the inside and phloem is towards the outside.

c. The intracellular spaces function in gas storage and exchange.

p. 28-15 **Figure 28-8** *Cross section of a dicot leaf*
(clockwise from right): palisade mesophyll; spongy mesophyll; stomate.

C   Monocot Leaf Structure—Corn (*Zea mays*)

d. This mechanism conserves water by reducing the amount of surface exposed to desiccation.

e. Both. No, there are more on the lower epidermis.

f. In dicots, stomates are only on the lower epidermis. This arrangement is related to the leaf and petiole configuration of dicot versus monocot leaves. Dicot leaves are petiolate (have petioles) and therefore lie flat, with the lower surface shaded from the sun. Monocot leaves often lack petioles and 'clasp' the stem, Therefore, the lower and upper epidermal surfaces are almost equally exposed.

g. Corn lacks palisade mesophyll; corn contains only spongy mesophyll.

h. The xylem is located above the phloem (closer to the upper surface of the leaf).

p. 28-16   Figure 28B-9 *Cross Section of a monocot leaf (corn,* Zea mays*)*
(a) upper epidermis; (b) stomate; (c) mesophyll; (d) xylem; (e) phloem; (f) bundle sheath cells; (unlabeled line) lower epidermis.

Exercise C   The Pine Leaf (Optional)

p. 28-16   Figure 28C-1 *Cross section of a pine leaf*
(a) epidermal cuticle; (b) hypodermis; (c) resin ducts; (d) endodermis; (e) vascular bundles; (f) transfusion tissue; (g) mesophyll; (h) stomate.

Laboratory Review Questions and Problems

Summary of Tissue Systems, Tissues, and Cell Types

Tissue systems	Tissues	Cell types	Cell functions
Dermal	Epidermis	parenchyma (guard cells, trichomes; sclerenchyma)	protection (gas exchange, water and mineral absorption, other metabolic functions)
	Periderm	parenchyma (sclerenchyma)	cork protective; cork cambium produces cork
Vascular	Xylem	tracheids, vessel members	conduction of water
	Phloem	sieve cells; sieve tube members, companion cells	food conducting; moving food into and out of sieve cells
Ground (or fundamental)	Parenchyma	parenchyma cells (thin walled)	most abundant cell type; packing tissue, metabolism, wound healing, regeneration
	Collenchyma	collenchyma cells (elongate, irregular with thickened cell wall)	support in primary cell body
	Sclerenchyma	fibers, sclerids	support

1. Dermal tissue is the outer covering of a plant. Dermal cells are specialized for protection and may be structurally modified (hairs, trichomes, stomates, cuticles) to retard water loss. Vascular tissue contains the conducting cells (xylem and phloem). Ground tissue consists of cells specialized for food storage and support (parenchyma, sclerenchyma, and collenchyma).

2. Epidermis—The outermost layer of cells of a young root or leaf; primary tissue in origin.
   Periderm—outer protective tissue that replaces the epidermis when it is destroyed during secondary growth; includes cork and cork cambium.
   The epidermis usually consists of parenchyma cells, guard cells, and trichomes, as well as some sclerenchyma cells. The periderm is composed of parenchyma cells and some sclerenchyma cells.

3. Xylem cells are dead at maturity. The lignified secondary walls function in water transport. Phloem cells have only primary cell walls (no secondary walls), are alive at maturity, and function in transport of photosynthate. Both are vascular (transport) tissues.

4. All are components of the ground tissue. Parenchyma is thin walled, alive at maturity and

often stores starch. Collenchyma has thickened primary walls, is alive at maturity and provides flexible support. Sclerenchyma is usually dead at maturity, has thickened, often lignified secondary walls, and provides rigid support.

5. Cortex and pith are composed of parenchyma, and are often sites of starch storage. A stele is the region of a root inside the endodermis. It consists of pericycle; primary phloem; primary xylem; and vascular cambium. The endodermis is the innermost layer of the cortex and is characterized by the Casparian strip (a waxy band around each cell).

6. Most monocots are herbaceous and lack secondary growth (although some, such as palms, can increase girth by primary growth). Dicots can be either herbaceous or woody. Woody dicots exhibit secondary growth as the result of activity by the vascular cambium and cork cambium.

7. Structural differences that can be used to distinguish monocots and dicots include:

Dicots	Monocots
flower parts usually in fours or fives	flower parts usually in threes
seeds have 2 cotyledons	seeds have 1 cotyledon
netlike leaf venation	parallel leaf venation
primary vascular bundles in a ring in the stem	primary vascular bundles dispersed in stem
secondary growth with vascular cambium common	secondary growth absent
three furrows or pores in pollen grains (tricolpate)	one furrow or pore in pollen grains (monocolpate)

8. Three functions of roots include: 1) food storage (e.g., carrots, sweet potatoes); 2) anchoring; and 3) absorption of water and minerals.

9. Three functions of stems include: 1) support of photosynthetic structures (leaves); 2) transport; 3) food storage (e.g., white potatoes are underground stems)

10.

	Arrangement and location of vascular tissues	Presence or absence of stele	Presence or absence of pith	Special characteristics
Dicot stem	vascular bundles arranged in a ring around a pith	no	yes	secondary growth
Monocot stem	vascular bundles scattered	no	no	no secondary growth
Dicot root	central xylem and phloem patches between xylem arms	yes	no	endodermis
Monocot root	alternating patches of xylem and phloem arranged in a cylinder	yes	yes	endodermis
Dicot leaf	netted venation (xylem = upper; phloem = lower)	no	no	palisade and spongy mesophyll
Monocot leaf	parallel venation (xylem = upper; phloem = lower)	no	no	spongy mesophyll only

11. In C_4 plants, CO_2 is first fixed in the mesophyll cells by binding to physophoenolpyruvate (PEP). The enzyme PEP carboxylase is repsonsible for this reaction that results in formation of oxaloacetic acid (OAA). The OAA is reduced by $NADPH + H^+$ to malic acid which is then transported to bundle sheath cells. The bundle sheath cells are parenchyma (or sclerenchyma) cells that surround vascular bundles in the leaf. The Calvin cycle takes place in the bundle sheath cells of C_4 plants. When CO_2 enters the Calvin cycle, pyruvic acid (malic acid + $NADP^+ \rightarrow CO_2$ + pyruvic acid + $NADPH + H^+$) is returned to the mesophyll cells.

12. In simple leaves, the blade is single while in compound leaves, the blade is divided into leaflets. If a leaf is palmately compound, all of the leaflets originate from a single point on the petiole. If a leaf is pinnately compound, the leaflets arise from many different points on the petiole in even or odd patterns. Leaves with netted venation have a pattern of highly branched veins (vascular tissue). This is typical of dicots. Leaves with parallel venation show little or no branching, with most vascular tissue running in parallel lines. This pattern is typical of monocots.

Notes:

Laboratory 29

Angiosperm Development—Fruit, Seeds, Meristems, and Secondary Growth

EXERCISE A   Fruits

p. 29-4   Table 29A-2 Identification of Common Fruits

Plant (common name)	FRUIT Q	Portion Eaten (if applicable)	Means of Seed Dispersal
1. coconut	drupe	seed -endocarp	floats in ocean
2. orange	hesperidium	endocarp	animal
3. apple	pome	floral tube	animal
4. tomato	berry	pericarp	animal
5. banana	berry	mesocarp and endocarp	animal
6. grape	berry	mesocarp and endocarp	animal
7. strawberry	aggregate	receptacle	animal
8. fig	multiple	receptacle	animal
9. pineapple	multiple	stem and flower	animal
10. corn	grain	pericarp and seed coat	animal
11. milkweed	follicle	not eaten	wind

EXERCISE B   Seed Structure

PART 1   Examining the Dicot Bean Seed

p. 29-5(6) Figure 29B-1 (a) The bean seed, a dicot

(a) pericarp; (b) endosperm; (c) cotyledon; (d) coleoptile; (e) epicotyl; (f) first leaves or plumule.

PART 2   Examining the Monocot Corn Seed

p. 29-5(6) Figure 29B-1 (b) A seed of corn, a monocot

(a) pericarp; (b) endosperm; (c) cotyledon; (d) coleoptile; (e) epicotyl; (f) hypocotyl; (g) coleorhiza.

  a. In dicots such as bean, the two cotyledons are the main food storage organs for the germinating seed. Dicot cotyledons acquire food reserves from digestion of endosperm during seed development. In contrast, monocot seeds store food as endospem.

  b. The single cotyledon (scutellum) of corn absorbs the food reserves of the endosperm for use during seed germination.

EXERCISE C   Found a Peanut

p. 29-6   **a.** Legume.
b. No. The peanut is not a nut because it is dehiscent. Nuts are indehiscent (do not open at maturity). The peanut is a fruit.
c. The shell of a peanut is the pericarp.
d. Two.
e. Seed coat.
f. Cotyledons and embryo.
g. Dicot.
h. Cotyledons.
i. The leafy structures are vascularized. Cotyledons are specialized for food storage and are not vascularized.

EXERCISE D   Seedling Development

PART 1   Comparing Germination in Beans, Peas, and Corn

p. 29-7   **a.** The hypocotyl arch protects the coyledons and the apical meristem as it pulls them up through the soil.
b. The hypocotyl becomes the stem.

p. 29-8   **c.** Ungerminated seeds are positive for I_2KI.
d. The food reserves in the germinated seedling were used for growth.

p. 29-9   **Table 29D-1 Above and Below Ground Structures**

	Bean	**Pea**	**Corn**
Above ground	cotyledons, epicotyl, part of hypocotyl	epicotyl shoot	coleoptile
Below ground	primary root hypocoytl	cotyledons, cotyledon, root	coleorhiza

PART 2   Observing the Germination and Development of Seeds

p. 29-9   **Table 29D-2 Seed Germination**

Type of Seed	Time until Emergence of Epicotyl	Time until Emergence of Hypocotyl	Fate of Cotyledons	Fate of Hypocotyl and Epicotyl	Day 5: Length of Hypocotyl + General Appearance	Day 10: Length of Root and Shoot + General Appearance
Bean	5 days	2 days	Wither as food supply is exhausted and dehisce	Stem	hypo: 6 cm epi: 1.8 cm Two true leaves	root: 9 cm shoot: 18 cm Mature plant; cotyledons exhausted
Corn	2 days	(radicle) 1 day	Remain below ground	Epcotyl becomes shoot; hypocotyl becomes primary root	hypo: 2.4 cm cep: 4.9 cm First leaf	root: 6 cm shoot: 5 cm (stem alone); 11 cm with leaves) Mature plant
Pea	2 days	2 days	Remain below ground	Epicotyl becomes shoot; hypocotyl becomes root	hypo: 1.5 cm epi: 4 cm Several true leaves	root: 8 cm shoot: 15 cm Mature plant

EXERCISE E   Studying the Stem Tip and Root Tip

PART 1   Examining the Stem Tip

p. 29-10   Figure 29E-1 Stem tip of the *Coleus* plant
(a) meristematic cells; (b) leaf primordia; (c) vascular tissue; (d) protoderm; (e) procambium; (f) ground meristem.

p. 29-11   a. Opposite.
b. floral inflorescence

PART 2   Examining the Root Tip

p. 29-11   Figure 29E-2 Root tip
(a) root cap; (b) apical meristem; (c) region of cell division; (d) region of elongation; (f) ground meristem; (g) protoderm

Extending Your Investigation: Growing Longer

p. 29-12   HYPOTHESIS: If most mitosis takes place in the apical meristem of the root tip, it will increase in length faster than other areas of the root tip.

NULL HYPOTHESIS: All areas of the root tip will increase in length by the same amount as a pea seed germinates.

Independent variable—time.

Dependent variable—length of root section.

1.   *Pea seedling*   *Pea seedling after 24 hours*

   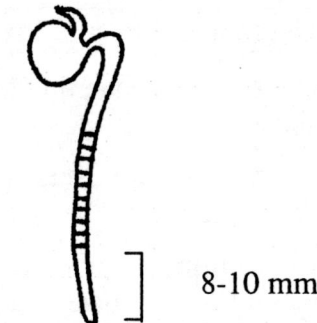

8-10 mm

p. 29-13   RESULTS:

Data do not support the hypothesis. Most growth took place in the area behind the apical meristem in the region of elongation.

Observations allow the null hypothesis to be rejected.

The prediction was incorrect.

Conclusion—Although mitosis leads to formation of new cells, most growth is in the area posterior to the mitotic portion of the root tip. This area is the area of elongation.

Table 29E-1 Root Length of Germinating Pea Seedlings

Pea Seedling	Number of Marks 2 mm Apart	Total Length of Root at Start (mm)	Root length after 24 hours	
			Distances Between Marks	Total Growth (mm)
1	8	20	8 mm tip to mark 1	9
2	10	25	10 mm tip to mark 1	10
3	8	22	10 mm tip to mark 1	

EXERCISE F   Secondary Growth of Angiosperms—The Woody Stem (*Tilia*)

p. 29-14   **Figure 29F-2 Cross section of *Tilia* stem**
(a) pith; (b) secondary xylem ("wood"); (c) primary xylem; (d) vascular cambium; (e) vascular ray; (f) annual ring; (g) spring wood; (h) summer wood; (i) phloem; (j). phloem fibers; (k) sieve tubes and companion cells;

p. 29-15   (l) cortex; (m) cork; (n) epidermis
  a. 35-50 years old (count the rings; student answers will vary)
  b. No, or are less distinct
  c. Summer xylem is located to the inside. It is the end of the previous year's growth.

p. 29-16   **d.** The bark includes the phloem, cortex, cork and epidermis. Wood is the secondary xylem.

EXERCISE G   Plant Tissue Culture

Laboratory Review Questions and Problems

1. A fruit is a mature seed-containing ovary, a cluster of mature ovaries, or an ovary and closely associated tissues.

2. A peanut is a dehiscent fruit. Nuts are indehiscent fruits. Therefore, a peanut should not be called a nut. It should simply be called a fruit.

3. Simple fruits arise from the ovary of a single flower. Multiple fruits consist of enlarged ovaries of several flowers coalesced into one mass. Aggregate fruits are composed of multiple ovaries of a single flower scattered over the surface of a single receptacle.

4. Fleshy fruits have a thickened peripcarp while dry fruits do not. Dry fruits may split open at maturity (dehiscent) or not (indehiscent).

5.

Fruit	Simple, multiple, or aggregate	Type of simple fruit
Plum	simple	drupe
Strawberry	aggregate	achene
Peanut	simple	legume
Pineapple	multiple	berry
Banana	simple	berry
Fig	multiple	drupe
Tomato	simple	berry
Coconut	simple	drupe
Orange	simple	hesperidium

6. Parents should say, "Eat your fruits!" Vegetables are fruits.

7.

"Vegetable"	Plant Part Eaten	"Vegetable"	Plant Part Eaten
Carrot	root	Spinach	leaves
Irish Potato	root	Squash	fruit
Sweet Potato	modified stem	Lettuce	leaves
Celery	stem	Onion	leaves
Broccoli	flowers and stem	Corn	seed

8.

		Monocot	Dicot
Seed	Number of cotyledons	1	2
	Presence of plumule	no	yes
	Presence of coleoptile	yes	no
	Presence of coleorhiza	yes	no
Stem	Presence of vascular cambium	no	yes
	Presence of primary meristem	yes	yes

9. **fruit, ovary:** The ovary wall forms the pericarp of the fruit.

   seed, ovule: A seed is a matured ovule.

   indehiscent, dehiscent: Dehiscent fruits open at maturity. Indehiscent fruits do not open.

   hypocotyl, epicotyl: The parts of the plant embryo located above the plane of attachment of the cotyledons constitute the epicotyl. Embryo parts below this plane constitute the hypocotyl.

   monocot, dicot: The two groups of angiosperms, distinguished by characteristics of: (1) leaf venation (monocots = parallel; dicots = netted); (2) the number of flower parts (monocots in threes; dicots in fours or fives); (3) lack of vascular cambium (lack of secondary growth) in monocots; (4) vascular tissue arrangement in stems and roots.

   branch roots, adventitious roots: Branch roots are lateral roots which develop on the primary root and arise in the pericycle of the root. Adventitious roots develop from stem nodes.

   primary meristem, secondary meristem: The root and shoot apical meristems are primary meristems. They increase the length of a plant. Secondary meristems, such as the vascular cambium and cork cambium, increase the width of a stem.

   wood, secondary xylem: Wood is composed of secondary xylem.

   spring wood, summer wood: Both are composed of secondary xylem vessels. The vessels in spring wood are larger in diameter than are the vessels in summer wood.

   heartwood, sapwood: Both are composed of secondary xylem. Heartwood is impregnated with tannins and does not function as a transport pathway. Sapwood is functional secondary xylem.

   bark, cork: Cork is produced by the cork cambium. Bark is composed of all tissues outside of the vascular cambium. Bark includes the secondary phloem and the cork.

10. The bark of a tree is composed of the secondary phloem and the cork. "Girdling" a tree kills the tree because the phloem conducting tissue is located within the bark. Therefore, cutting completely through the bark disrupts the flow of photosynthate through the phloem. Once phloem translocation is disrupted, sucrose produced in the leaves can no longer supply the plant. Thus the tree dies due to a lack of organic nutrients.

Notes:

Laboratory 30

Water Movement and Mineral Nutrition in Plants

EXERCISE A   Observing Stomata

p. 30-2   Table 30A-1   Distribution of Stomata

Name of Plant Used:	Zebrina		Rhoeo		Elodea	
	lower epidermis	upper epidermis	lower epidermis	upper epidermis	lower epidermis	upper epidermis
Number of stomata in one field of view	36	0	40	0	0	0
Average	40	0	38	0	0	0

    a. Yes. Stomata are more numerous on the lower leaf surface.

    b. (see above chart). Plants adapted to different habitats have different numbers of stomata on the upper and lower epidermis. For example, the submersed aquatic plant, *Elodea,* lacks stomata, both *Rhoeo* and *Zebrina* have abundant stomata on the lower epidermis, but none on the upper epidermis.

p. 30-3   **c.** Gas exchange occurs directly through the epidermal cells. Submersed aquatic plants, such as *Elodea,* either lack cuticle or have very thin cuticles.

    d. Yes.

    e. Closed.

    f. Guard cells usually take up water when their solute content increases and water enters. The ends of the swelling guard cells push against each other and the stoma opens. When water leaves the leaf cells because they are placed in sucrose (high $\Psi_{leaf} \rightarrow$ low $\Psi_{sucrose}$) the guard cells become flaccid.

    g. Flattened.

    h. Yes. Closed, kidney-shaped, close together.

    i. Because water left the guard cells

EXERCISE B   Guttation

p. 30-3   **a.** At the tips of the leaves

EXERCISE C   Transpiration

PART 1

p. 30-4   **a.** Water droplets formed in Flask 2, which contained the *Coleus* shoot.

    b. Flask 1. This flask monitors the moisture content of the air in the absence of

Coleus plant transpiration.

    c. The water comes from the mesophyll cell of the leaves, exiting leaves through the stomata.

    d. The petroleum jelly served to occlude stomatal openings and effectively block all evaporation from leaf surfaces.

PART 2

p. 30-5   HYPOTHESIS: Heat and wind will increase transpiration but increased humidity will decrease transpiration.

NULL HYPOTHESIS: Transpiration rates are constant and unaffected by environmental variables.

Prediction—transpiration rates are unaffected by environmental conditions.
Independent variable—environmental conditions
Dependent variable—transpiration rates

p. 30-6   **Table 30C-1 Transpiration Experiment Data**

Readings	Room Conditions Water loss per 10-minute interval	Readings	Experimental Condition (+ fan) Water loss per 10-minute interval
a ml at start 0 **b** ml at 10 min. 0.11 **c** ml at 20 min. 0.21	**b − a** 0.11 **c − b** 0.10 Average water loss per 10-minute interval 0.105	A ml at start 0.11 B ml at 10 min. 0.25 C ml at 20 min. 0.35	**B − A** 0.14 **C − B** 0.10 Average water loss per 10-minute interval 0.12

Readings	Room Conditions Water loss per 10-minute interval	Readings	Experimental (+ high vapor pressure) Water loss per 10-minute interval
a ml at start 0.14 **b** ml at 10 min. 0.24 **c** ml at 20 min. 0.34	**b − a** 0.10 **c − b** 0.10 Average water loss per 10-minute interval 0.10	A ml at start 0.09 B ml at 10 min. 0.13 C ml at 20 min. 0.18	**B − A** 0.04 **C − B** 0.05 Average water loss per 10-minute interval 0.045

Readings	Room Conditions Water loss per 10-minute interval	Readings	Experimental Condition (+ heat + light) Water loss per 10-minute interval
a ml at start 0.06 **b** ml at 10 min. 0.17 **c** ml at 20 min. 0.27	**b − a** 0.11 **c − b** 0.10 Average water loss per 10-minute interval 0.105	A ml at start 0.17 B ml at 10 min. 0.31 C ml at 20 min. 0.48	**B − A** 0.14 **C − B** 0.17 Average water loss per 10-minute interval 0.155

Results support the hypothesis: high heat increased transpiration rate the most (however, this is not always the case). Depending on the relationship of the plane of the leaves to the wind coming from the fan, wind sheer can cause stomata to close and lower the rate of transpiration. If this does not happen, the transpiration rate under windy conditions can be higher than the rate in high heat conditions.

Results allow the null hypothesis to be rejected.

Prediction—correct.

Conclusion—environmental conditions affect transpiration rate.

   a. Transpiration was greatest under conditions of increased light and heat.

   b. Wind diminishes the vapor pressure near the leaf surfaces and therefore increases transpiration rate. High vapor pressure (100% humidity environment) decreases the rate of transpiration.

p. 30-7   **Figure 30C-2** *Class data for effects of environmental conditions on transpiration rate*

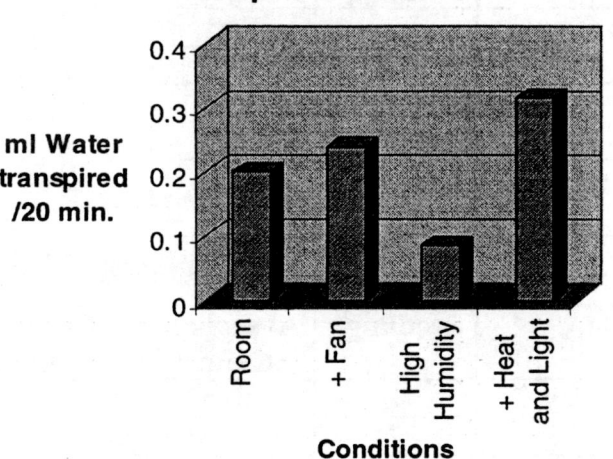

   c. Yes. Transpiration rate will depend upon the number and size of the leaves and the metabolic state or age of the leaves.

   d. Yes. Larger leaves increase transpiration rate.

   e. Transpiration is increased by a greater number of stomata, especially a greater number on the upper leaf surface that is exposed to the sun.

EXERCISE D   The Pathway of Water Movement Through a Plant

p. 30-7   **a.** Yes

p. 30-8   **b.** Xylem

EXERCISE E   Plant Mineral Nutrition

p. 30-9   HYPOTHESIS: Since nitrogen is a necessary macronutrient for plant growth, its absence in hydroponics medium will result in plants that show stunted growth and less biomass as well as a lower chlorophyll content.

   NULL HYPOTHESIS: Levels of nitrogen will not make a difference in plant growth.

P. 30-10   Prediction—Plants grown without nitrogen will be have less biomass and chlorophyll than plants grown with necessary macro- and micronutrients.
Independent variable—levels of macro or micronutrients.
Dependent variable—growth (height and leaf spread), biomass and chlorophyll content.

p. 30-10   Table 30E-2 Data on Seedling Mass (–N)

Week	Seedling	Mass (g)
Time zero (day 1)	1	1.04
	2	0.99
1 week growth	1	1.80
	2	1.85
2 weeks growth	1	1.99
	2	2.01

Table 30E-2 Data on Seedling Mass (complete)

Week	Seedling	Mass (g)
Time zero (day 1)	1	1.03
	2	1.04
1 week growth	1	2.12
	2	2.00
2 weeks growth	1	2.61
	2	2.58

p. 30-11   Table 30E-3 Data on Chlorophyll Content (–N)

Week	Seedling	Absorbance (663 nm)	Chlorophyll Content (mg/l)	Standardized Chlorophyll (mg/g)
Time zero (day 1)	1	0.57	7.64	0.220
	2	0.63	8.48	0.257
1 week growth	1	1.10	14.74	0.246
	2	1.20	16.08	0.260
2 weeks growth	1	0.63	8.45	0.127
	2	0.66	8.84	0.132

Table 30E-3 Data on Chlorophyll Content (complete)

Week	Seedling	Absorbance (663 nm)	Chlorophyll Content (mg/l)	Standardized Chlorophyll (mg/g)
Time zero (day 1)	1	0.77	10.32	0.30
	2	0.70	9.38	0.27
1 week growth	1	1.30	17.66	0.25
	2	0.80	10.67	0.16
2 weeks growth	1	1.25	16.75	0.19
	2	1.30	17.42	0.20

Figure 30E-1 *Biomass of plants grown on complete vs. nitrogen deficient medium*

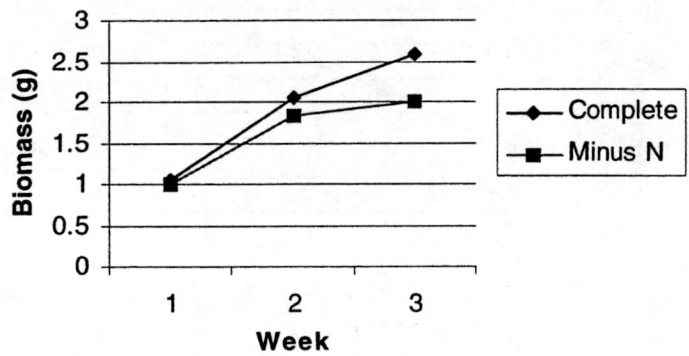

Figure 30E-2 *Chlorophyll content of plants grown on complete vs. nitrogen deficient medium*

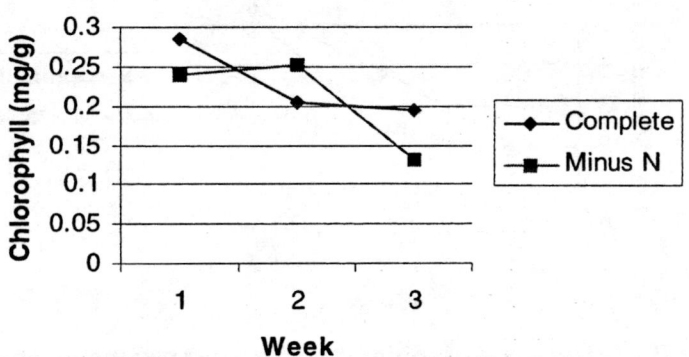

Table 30E Height and Leaf Spread of Plants (-N)

Week	Seedling	Height	Leaf Spread
1	1	2.6	1.8
	2	3.0	2.0
2	1	4.8	2.2
	2	4.5	2.5
3	1	6.2	4.8
	2	6.0	5.5

Table 30E Height and Leaf Spread of Plants (complete)

Week	Seedling	Height	Leaf Spread
1	1	2.67	1.9
	2	2.80	2.2
2	1	4.25	3.4
	2	5.40	4.2
3	1	6.50	5.2
	2	6.25	5.5

Figure 30E-3 *Height of plants grown in complete vs. nitrogen-deficient medium*

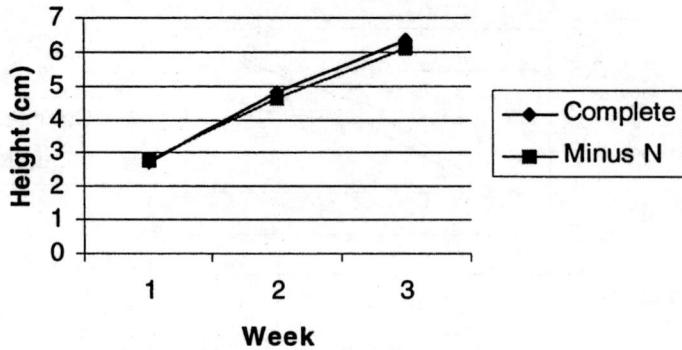

Figure 30E *Leaf spread of plants grown in complete vs. nitrogen deficient medium*

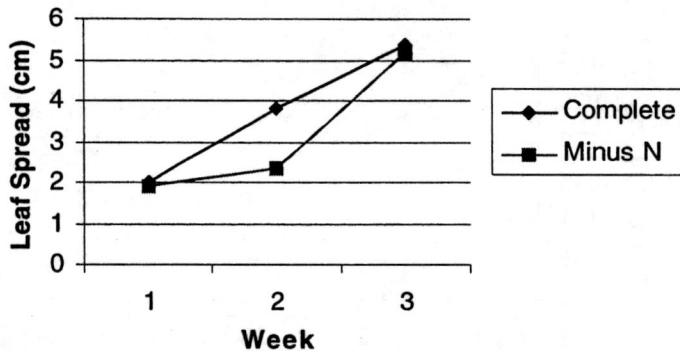

Note: If observations of mineral nutrition experiments are continued for at least 5 weeks, results of mineral deficiencies are very obvious (see below). Deficiencies in the first few weeks are not pronounced since most food for growth comes from the cotyledons of the plants. (See *Preparator's Guide* for suggestions on how to set up the experiment to obtain best results.)

Figure 30E *Height of plants grown in complete vs. nitrogen-deficient medium*

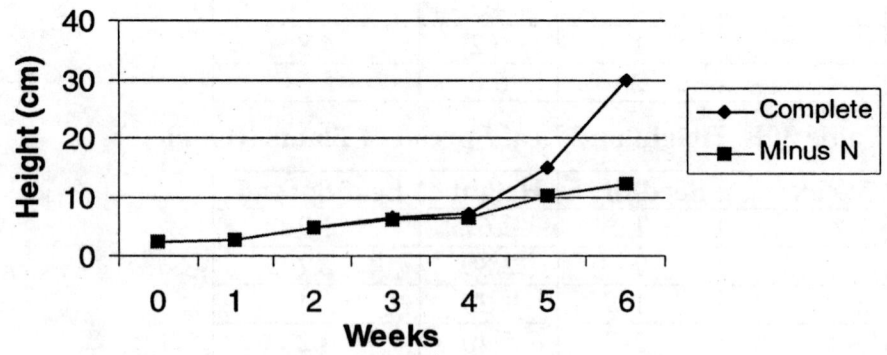

Class data were collected from all sections of students as shown below for the following treatments: complete, –N, –K, –P, and –Fe.

Table 30E Week One Data

Treat-ment	Average Biomass (g)	Average Standardized Chlorophyll Content (mg/g)	Maximum Biomass (g)	Maximum Standardized Chlorophyll Content (mg/g)	Minimum Biomass (g)	Minimum Standardized Chlorophyll Content (mg/g)
complete	1.03	0.290	1.68	1.165	0.34	0.060
–N	0.98	0.220	1.60	0.363	0.59	0.098
–K	1.03	0.221	1.47	0.368	0.63	0.054
–P	1.01	0.208	1.70	0.431	0.41	0.063
–Fe	0.92	0.238	1.53	0.370	0.47	0.064

Table 30E Week Two Data

Treat-ment	Average Biomass (g)	Average Standardized Chlorophyll Content (mg/g)	Maximum Biomass (g)	Maximum Standardized Chlorophyll Content (mg/g)	Minimum Biomass (g)	Minimum Standardized Chlorophyll Content (mg/g)
complete	2.13	0.230	6.01	0.543	0.69	0.094
–N	1.88	0.247	3.35	1.820	0.68	0.079
–K	1.71	0.281	2.50	0.499	0.55	0.150
–P	1.72	0.215	3.05	0.423	0.76	0.091
–Fe	2.50	0.141	4.24	0.250	0.95	0.009

Table 30E Week Three Data

Treat-ment	Average Biomass (g)	Average Standardized Chlorophyll Content (mg/g)	Maximum Biomass (g)	Maximum Standardized Chlorophyll Content (mg/g)	Minimum Biomass (g)	Minimum Standardized Chlorophyll Content (mg/g)
complete	2.66	0.191	5.28	0.402	0.51	0.069
–N	2.03	0.128	3.69	0.402	0.14	0.026
–K	2.73	0.212	5.55	0.354	0.82	0.080
–P	2.63	0.157	6.56	0.359	0.67	0.072
–Fe	3.39	0.113	8.71	0.287	1.47	0.009

Chi square data calculated using the statistics package on the accompanying CD-ROM indicated the following ($df = 4$):

Biomass

	Week 1	Week 2	Week3
Chi-square	2.33	10.63	6.57
Probability	50-75%	2.5-5%*	10-25%

Standardized Chlorophyll

	Week 1	Week 2	Week3
Chi square	38.41	8.33	35.53
Probability	<0.5%**	5-10%	<0.5%**

Differences in chlorophyll content were found to be highly significant (**) in weeks 1 and 3. Differences in biomass were only significant in week 2.

Figure 30E *The effects of mineral nutrition on biomass of growing plants*

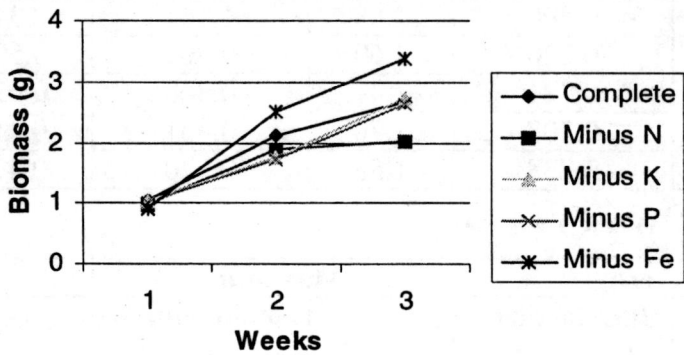

Figure 30E *The effects of mineral nutrition on standardized chlorophyll content of growing plants*

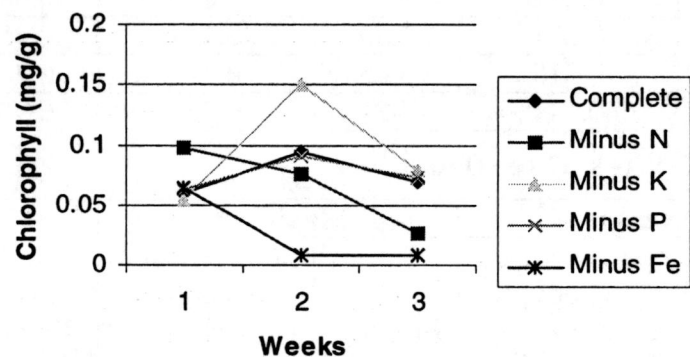

Table 30E Class Data Height (cm)

Treatment	Week 1	Week 2	Week 3	Δ
complete	2.4	4.9	6.4	4.0
–Fe	3.4	3.6	5.8	2.4
–N	2.8	4.5	6.1	3.3
–Ca	3.2	3.5	5.8	2.6
–K	3.7	4.0	5.3	1.6
–P	2.5	3.5	5.6	3.1
–Mg	2.5	4.0	6.0	3.5
–S	2.8	3.6	5.8	3.0

Table 30E Class Data Leaf Spread (cm)

Treatment	Week 1	Week 2	Week 3	Δ
complete	2.0	3.8	5.4	3.4
–Fe	2.1	4.1	4.5	2.5
–N	1.9	2.4	5.1	3.2
–Ca	2.4	3.2	4.4	2.0
–K	1.2	3.5	4.0	2.8
–P	2.1	2.3	4.5	2.4
–Mg	2.1	4.2	4.5	2.4
–S	2.0	3.9	4.0	2.0

p. 30-12   Table 30E-4 Appearance of Plants after 1 and 2 Weeks (Weeks 2 and 3)

Week	Treatment	Observations
1 week growth	Control	
	– K	plants same size as others, upper leaves small
	– N	plants lighter green but large
	– Fe	plants large and growing well
	– P	plants medium in size and very green
	– Ca	plants smaller than most, some browning
	– Mg	plants lighter in color and not as tall
	– S	plants small; not as robust as most
2 weeks growth	Control	
	– K	margins of lower leaves are curled under and necrotic, especially at apex and margins
	– N	plants a lighter green than most
	– Fe	plants large but upper leaves yellowing
	– P	lower leaves are purplish, otherwise growth is robust
	– Ca	stunted growth or dead plants
	– Mg	lower leaves are dead, growth is stunted
	– S	necrotic, small leaves

Table 30E Mineral Deficiencies and General Plant Appearance

Mineral	Function	Appearance of Plant if Deficient
potassium	osmotic balance	younger (upper leaves) remove mobile potassium from the older (lower) leaves, destroying the osmotic balance in the lower leaves
nitrogen	constituent of nucleic acids, amino acids, chlorophyll, hormones	plant yellowish in color due to lack of chlorophyll, turns brown due to lack of photosynthesis; growth stunted; dies at later stages
iron	energy transfer molecules in photosynthesis and respiration	iron is relatively immobile, therefore iron in the lower leaves is not used to supply the upper leaves; upper leaves are yellow from lack of chlorophyll
phosphorus	constituent of nucleic acids; functions in energy metabolism	phosphorus is mobile, therefore is transported from older to younger leaves; older leaves accumulate anthocyanins, resulting in a purplish color
calcium	cell wall constituent	normal growth is impossible without cell wall synthesis
magnesium	part of chlorophyll molecule; enzyme activator	accelerated chlorosis of lower leaves occurs as the younger leaves remove mobile magnesium from the older leaves
sulfur	component of proteins and coenzyme	generalized growth inhibition due to lack of protein synthesis

a. Plants in complete medium showed a greater increase in biomass than those in –N medium. When all nutrients are considered, those in complete medium and those in –N and –Mg media showed the greatest increase in height. Those in complete media and media deficient in N or K showed the greatest leaf spread. Results could have been affected by the fact that the experiment was started before all the food and reserve nutrients available in cotyledons had been used up.

b. Yes, the –Ca plants began to turn yellow. The plants deficient in N were a lighter green and those deficient in P were beginning to darken on the bottom leaves

c. The –Ca plants had cell membranes and walls affected and probably could not manufacture chloroplasts. Fe^{2+} is used in electron transport and chlorophyll synthesis and lack of chlorophyll synthesis led to lighter leaves. Nitrogen is also a constituent of the porphyrin ring of chlorophyll and its absence would reduce chlorophyll synthesis. Phosphorus is mobile and is transported from older leaves to newer leaves. Anthocyanins then collect in the older leaves and they turn purple.

d. Yes. Growth was stunted in the –Ca plants and the –P and –K plants were not as bushy. Growth lags behind those in complete medium.

e. Plants in complete and –K media showed the greatest chlorophyll content. Potassium affects osmotic balance but is not necessary for chlorophyll synthesis. Since cells may not have contained as much water, the chlorophyll content (mg/g) may have been higher than chlorophyll content in plants incubated with complete medium.

f. Yes, there were differences in all experimental treatments compared to control treatments.

g. Differences in chloroplast content were found to be very significant or fairly

significant for all experimental plants and for the –N treatment. Differences in
biomas were only significant for week 2.

The prediction was correct; –N plants did not grow as well (biomass and spread) and
did not contain as much chlorophyll as those in complete medium

Results support the hypothesis and allow null hypothesis to be rejected.

Conclusion—both micronutrients and macronutrients are necessary for proper plant
growth.

Laboratory Review Questions and Problems

1. Continued transpiration of the cut flowers draws air (rather than water) into the xylem at the base of the stem. Therefore, even when placed into water, the stems cannot supply the leaves and stems with water because air bubbles break the cohesion of the water column within the xylem. Cutting the stems exposes a xylem surface that (may) still retain an uninterrupted cohesive column of water.

2. Transpiration is greater from leaves having greater surface areas (ie. larger leaves). Therefore, plants in moist, shady forests can have greater leaf surface areas, yet not transpire excessively. Plants in hot, dry environments must modify leaves (as by reduction of surface area) to reduce transpiration.

3. CO_2 is fixed in the photosynthetic Calvin-Benson Cycle. Active photosynthesis by mesophyll cells depletes the concentration of CO_2 within leaves, Therefore, stomata open in response to the low internal CO_2 levels, allowing atmospheric CO_2 to enter the cells.

4. Stomata close in response to high concentrations of CO_2 within the leaf. Rapid respiration produces high concentrations of cellular CO_2. Therefore, under hot, dry conditions, stomata remain closed. Closed stomata also conserve water (though water stress is secondary to CO_2 concentration in regulating stomata closing).

5. At night, CAM plants fix CO_2 into organic acids (malate) using PEP-carboxylase. Therefore, stomata open at night to admit CO_2. During the day, CAM plants keep stomata closed, but can fix CO_2 released from malate, using RUBP-carboxylase. CAM plants are adapted for environments having diurnally variable levels of atmospheric CO_2 (such as deserts) and often are favored by such environments which have hot days and cool nights.

6.

Modification	Environment	Advantage
Sunken stomata	Arid, windy	Reduces transpiration
Leaves modified as spines	Sunny, arid	Reduces water loss; also, plants living in sunny environments need less leaf area
Stomata only on upper epidermis	Float on water surface	Submerged leaves lack stomata and thick cuticles; floating leaves have stomata on upper surfaces to take advantage of the greater concentration and more rapid diffusion of CO_2 in air than in water

7. Plants deficient in magnesium or iron show variously localized yellowing or chlorosis (lack of chlorophyll) in the leaves, as demonstrated by the mineral nutrition experiment.

8. Nitrogen fixation is due to the presence of N2-fixing bacteria *(Rhizobium)* in the roots of

legumes. Rotation of plantings adds fixed nitrogen to the soil, especially if leguminous plants such as alfalfa are tilled under.

9. Mycorrhizal fungi extend the effective 'root area' of vascular plants, acting as would root hairs to absorb nutrients from the soil and transfer them to the cortical cells of the roots.

Laboratory 31
Plant Responses to Stimuli

EXERCISE A   Auxins

PART 1   Bud Inhibition and Apical Dominance

p. 31-2   HYPOTHESIS: FAA placed on the growing tips of stems will inhibit lateral bud development.

NULL HYPOTHESIS: FAA will have no affect on lateral bud development.

Prediction—plants treated with IAA will lengthen rather than growing bushier.

Independent variable—type of treatment: debudded, debudded + lanolin, debudded + lanolin + IAA.

Dependent variable—length of branches

Table 31A-1 Apical Dominance and Bud Inhibition

Treatment	Average Branch Length (mm)
Intact plants	17
Debudded plants	24
Debudded plants + lanolin	18
Debudded plants + lanolin and IAA	15

p. 31-3
   a. No
   b. Yes
   c. No
   d. Yes
   e. Lanolin serves as a control.

Conclusion—the substance produced by the apical bud and translocated down the stem to inhibit lateral growth is an auxin, such as IAA. In the absence of the apical bud (debudded plants) lateral growth is enhanced unless exogenous auxin IAA + lanolin treatment) is applied.

   f. Yes. Results support the hypothesis but lead to rejection of the null hypothesis.
   g. To compare experimental plants with plants normally producing auxin in apical buds.
   h. The effect is to encourage lateral growth and thereby cause plants to be "bushy."

PART 2   Leaf Abscission

p. 31-4   HYPOTHESIS: Auxin production by leaves prevents leaf abscission.

NULL HYPOTHESIS: Auxin has no effect on leaf abcission or hold.

Prediction—debladed petioles will drop if no auxin is applied.

Independent variable—treatment: lanolin, lanolin + IAA

Dependent variable—fate of leaf petiole

Table 31A-2 Leaf Abscission

Treatment	Observations
Debladed	petiole fell off
Debladed + lanolin	petiole fell off
Debladed + lanolin with IAA	petiole remained on plant

   a. Inhibits

   b. The auxin controlling the development of the abscission layer is produced in young tissues, such as growing leaves. Without the leaf, IAA was not produced for transport to the base of the petiole to retard leaf abscission.

   c. The lanolin treatments serve as controls. Since lanolin was used to apply IAA in the "Debladed + lanolin with IAA" treatment, the effects of lanolin alone on leaf abscission must be tested.

p. 31-5 Conclusion—auxin production by leaf blades is necessary to keep leaves from falling. When this production stops, the auxin levels at the base of the petiole are higher than those at the leaf and abscission occurs.

Results support the hypothesis and allow the null hypothesis to be rejected.

Exercise B   Gibberellins

p. 31-6   **Table 31B-1 Shoot Length in Response to Gibberellins**

	Initial Treatment: Stem Height (mm)	7 Days: Stem Height (mm)	Initial Treatment: Number OF Nodes	7 Days: Number OF Nodes
GA$_3$	7-9 mm	22-34 mm	2-3 mm	4-6 mm
Control	7-9 mm	11-14 mm	2-3 mm	5-6 mm

HYPOTHESIS: GA$_3$ will cause stem lengthening in dwarf pea plants.
NULL HYPOTHESIS: Adding GA$_3$ will have no effect on stem growth.
Prediction—Adding GA$_3$ will cause dwarf pea plants to "bolt."
Independent variable—GA$_3$ treatment.
Dependent variable—plant growth.

   a. Yes

   b. No

   c. Yes

Results support the hypothesis and allow the null hypothesis to be rejected.
Conclusion—GA$_3$ can cause dwarf pea plants to increase in shoot length.

   d. Dwarf plants produce less GA$_3$.

p. 31-7   **e.** Lack of the specific gene to produce GA$_3$ would cause dwarfism.

EXERCISE C   Tropisms

PART 1   Gravitropism (Geotropism)

p. 31-7   **a.** Stems are negatively gravitropic. Roots are positively gravitropic. Roots grow downward to a source of water and minerals while stems grow upward to lift their photosynthetic machinery to the light.

A   Negative Geotropism in Plants

p. 31-8   b. Auxins were concentrated in the lower surfaces of the horizontal plant.
      c. Cells on the lower side of the stem elongated, causing the stem to bend upwards.
      d. Plants were placed in the dark to eliminate light as a variable, since plants are positively phototropic

B   Positive Gravitropism in Roots: The Role of Seed Position

p. 31-8   HYPOTHESIS: The position in which a seed is planted affects the direction of root and stem growth.
       NULL HYPOTHESIS: The position in which a seed is planted does not affect the direction of root and stem growth.
       Prediction—roots will grow downward and stems upward no matter what direction the seeds are planted.
       Independent variable—position of seed.
       Dependent variable—direction of root growth.

p. 31-9   e. Despite the position of the seed, the ultimate orientation of the growing tip of the root is down (positively geotropic). To accomplish this orientation, roots growing from seeds which were placed upside-down or sideways curved downwards.
      f. Positive.
      g. This response ensures that the roots of a germinating seed will reach down into the soil for nourishment and stems will lift leaves toward light.
       Conclusion—seed position does not affect direction of root and stem growth during germination.
       Results support the null hypothesis and negate the alternative.
      h. Negative. The coleoptile grows in the opposite direction of gravity as do stems.

C   Positive Gravitropism in Roots: The Role of the Root Tip

p. 31-10   HYPOTHESIS: The root tip is responsible for the response of plant roots to gravity.
       NULL HYPOTHESIS: Positive gravitropism in roots is unaffected by the root tip.
       Prediction—roots without root tips will not bend downward.
       Independent variable—presence or absence of whole root tip.
       Dependent variable—exhibiting positive gravitropic response.

      i. Yes
      j. The roots of seedlings 2 and 3 curved downwards (if the root cap core was intact).
      k. Yes
       Conclusions—In horizontally oriented roots, the force of gravity is perceived by relocation of amyloplasts containing statoliths (starch grains) to positions near the lower walls of the root cap core cells. The accumulation of statoliths causes a localized increase in concentration of the growth inhibitor, abscisic acid (ABA). Inhibition of growth on these lower portions of cells causes the roots to bend downwards. Removal of the root cap prevents detection of gravity by the statoliths. Removal of part of the root cap does not completely restrict response to gravity, as long as part of the root cap core remains intact.
       Results support the hypothesis
       Results allow the null hypothesis to be rejected.

PART 2   Phototropism

p. 31-10   a. The actual bending occurs on the side of the stem away from the light.

b. Positive. This ensures that the major photosynthetic parts of the plant (leaves) are maximally exposed to light.

p. 31-11 **c.** Auxin concentration increased on the side of the stem away from the light, causing the stem to bend towards the light.

EXERCISE D   Light-Induced Germination

PART 1   The Role of Phytochrome

p. 31-12 **a.** Grand Rapids

b. No

c. Light-sensitive, Grand Rapids. Light-insensitive, Great Lakes.

Table 31D-1 Light Induced Germination in Lettuce Seeds

	Light		Dark	
Variety	**Grand Rapids**	**Great Lakes**	**Grand Rapids**	**Great Lakes**
Total number of seeds	50	50	50	50
Number germinated	39	33	15	33
Percent germinated	78%	66%	30%	66%

PART 2   The Forms of Phytochrome

p. 31-12   HYPOTHESIS: Red light will cause seeds to germinate.

NULL HYPOTHESIS: Color of light has no effect on seed germination.

Prediction—seeds will germinate best if exposed to red light.

Independent variable—color of light.

Dependent variable—number of seeds germinated.

p. 31-13   Table 31D-1 Effect of Wavelength of Light on Germination

	A	**B**	**C**	**D**	**E**	**F**
Light conditions	**Light**	**Dark**	**Blue**	**Green**	**Red**	**Far red**
Total number of seeds	50	50	50	50	50	50
Number germinated	39	15	7	17	41	0
Percent germinated	78%	30%	14%	34%	82%	0%

a. Red light converts phytochrome to the active form P_{fr} which induces germination.

b. Red and far red.

- Conclusion—germination was completely inhibited by far red light. Red light stimulates germination. Germination in blue or green light was not appreciably different from germination in the dark

Results support the hypothesis; far red light is included in the definition of "red" light.

Results allow the null hypothesis to be rejected.

c. Blue.

Figure 31D-1 *Effects of different wavelengths of light on inducing germination of lettuce seeds*

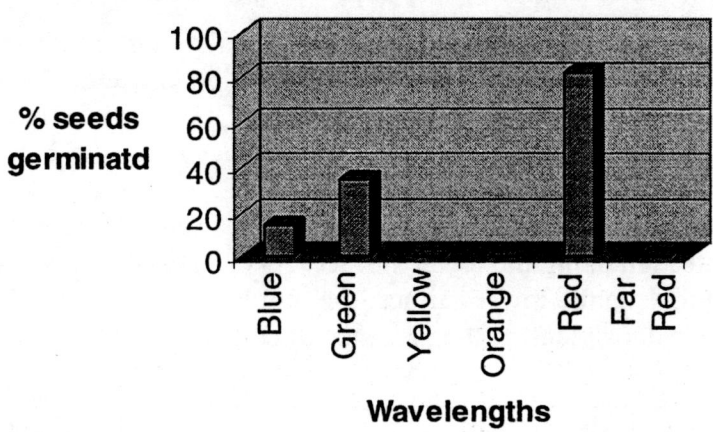

Effects of Light on Lettuce Germination

PART 3   Germination and Gibberellins (Optional)

p. 31-14   **Table 31D-3 Germination and Gibberellins**

Treatment	Percent Germination
H_2O/foil	15%
H_2O/light	100%
GA_3/foil	60-80%
GA_3/light	100%

a. Yes. More seeds germinate in the dark when gibberellic acid was used.
b. Gibberellic acid enhances the germination of light-sensitive seeds.

EXERCISE E   Photoperiodism

p. 31-15   **Table 31D-1 Photoperiodism**

Group	Photoperiod (hours) Light	Dark	Flowering (+ or −)	Category: long-day, short-day, or day-neutral
A (spinach)	8	16	−	long-day
	16	8	+	
B (petunia)	8	16	+	short-day
	16	8	−	
C (kidney bean)	8	16	+ or −	day-neutral
	16	8	+ or −	

a. A. P_{fr} promotes flowering: short night lessens $P_{fr} \rightarrow P_r$ conversion.
   B. P_{fr} inhibits flowering: long night allows $P_{fr} \rightarrow P_r$, so flowering occurs.
   C $P_{fr} \leftrightarrow P_r$ levels fluctuate without a specific effect on flowering.

1. External factors: light; temperature; nutrients; CO_2.
   Internal factors: solute concentrations; hormones; pigments; turgor pressure.

2. Hormones may either excite or inhibit specific plant functions.

3. When young monocot stems are oriented horizontally, auxin migrates to the lower side of the coleoptile and Ca^{2+} accumulates along the upper side. Since Ca^{2+} inhibits growth and auxin causes cells to elongate, the shoot bends upward. Once vertical, these differences disappear and the plant continues to grow upward. Whether a similar mechanism works in dicots is still not known for sure.

   Growth of horizontal roots in a downward direction is thought to be due to increased auxin concentration along the lower surface. As cells on the upper surface elongate more rapidly than those on the lower surface, the root bends downward. Transport of auxin to the lower surface is dependent on the presence of calcium.

4.

Hormones	Mechanism	Plant Response
Auxins	1. Production in roots and stems	1. Gravitropism
	2. Production in apical bud	2. Stem elongation
	3. Change in distribution in leaves or developing fruit	3. Leaf or fruit drop
	4. Light influences auxin distribution	4. Phototropism
Gibberellins	1. Synthesis in meristematic tissues	1. Stem lengthening, bolting
	2. GA_3 substitutes for effects of red light to break dormancy	2. Seed germination

5.

Hormone(s)	Mechanism	Plant Response
Cytokinins	1. Stimulate cell division	1. Shoot growth / increase in callus tissue
Ethylene	1. Production by fruits	1. Fruit ripening
	2. Production in abscission zone	2. Leaf and fruit abscission of leaf and fruit petioles
Abscissic acid	1. Stomatal closing in response to water stress	1. Prevents water loss from leaves
	2. Growth inhibition	2. Accelerates abscission

6. Cytokinins can reverse the inhibition of lateral bud growth that is produced by auxin from the apical bud. Auxin inhibits leaf abscission if applied before leaf senescence begins but promotes abscission (once the abscission layer has formed), by stimulating ethylene production. This induction of ethylene production by auxin also affects flower development and fruit ripening.

7. Phytochrome has two forms. It is synhesized in the P_r form and is converted to the P_{fr} form in daylight due to a reaction with a photon of red light. During the day the $P_r \rightarrow P_{fr}$ reaction reaches equilibrium. At night, P_{fr} levels decline as P_{fr} is converted to P_r. P_{fr} inhibits flowering in short day plants. But when the night is long enough, most P_{fr} is converted back into Pr and the plants can flower. In long day plants, P_{fr} promotes flowering; the night is short enough so that enough P_{fr} remains to promote flowering.

8. The plants in Group A `are short-day plants, since they flower when exposed to the 8 hours light/16 hours dark cycle. Based upon the evidence in the initial experiment, the plants in Group B could be either long day or day neutral plants. Interruption of the dark period by hourly flashes of red light effectively alters the photoperiod to a long day (short night) cycle, since plants actually measure the length of the dark period, rather than the length of the light period. The plants in Group A did not flower under the interrupted night cycle since they require long nights to flower. Since the plants in Group B did flower under this short night (long day) cycle, these plants must be long day plants.

More:

Laboratory 32

Animal Tissues

EXERCISE A   Epithelial Tissue

PART 1   Squamous Epithelium

p. 32-2   **a.** Polygonal.
           b. In sheets.

p. 32-3   **c.** Several layers of cells are present. In cross-section on a prepared slide, the cells appear to be stratified squamous epithelium but the outer layers are not strongly flattened and keratinized as in reptiles, birds, and mammals.
           d. Shed skin is epidermis only and chromatophores are found in the dermis. (In birds and mammals, portions of chormatophores (melanosomes) may enter epidermal cells and structures (feathers, hair.)
           e. Chromatophores are smaller and may have protoplasmic extensions. In living amphibians and reptiles chromatophores are capable of concentrating pigment (lightening the skin) or dispersing pigment (darkening the skin).
           f. Yes. In fixed tissues, cilia form mats on the apical edges of epithelia.
           g. Not appreciably.

PART 2   Columnar Epithelium

p. 32-3   **a.** Cilia in the reproductive system assist in the movement of gametes.
           b. Cilia in the respiratory system remove mucous with trapped foreign particles from the system.

p. 32-4   **c.** They are higher (from basal to apical ends) than they are wide—roughly columnar.
           d. Cells are arranged next to each other with the cell axis perpendicular to the lumen (simple epithelium).

PART 3   Cuboidal Epithelium

p. 32-4   **a.** Cells are arranged next to each other (simple epithelium).

p. 32-5   **b.** Yes.
           c. The particles move several millimeters in 1 minute (distance depends on body temperature).
           d. Yes.
           e. Yes.
           f. The cilia beat rhythmically in waves.

EXERCISE B   Connective Tissue

PART 1   Cartilage

p. 32-6   **a.** Elastic
           b. No (but fibers may be visible, dependent upon staining). Yes.

PART 2 Bone

p. 32-7 **a.** Vessels hemorrhage and form a hematoma (large clot). Nearby bone cells die and the tissue becomes swollen and inflamed. Over the next 3-4 weeks, a soft callus forms (capillaries grow into the hematoma and phagocytic white blood cells clear the remaining debris. Fibroblasts produce collagen fibers and osteoblasts begin to form tabeculae of new bone, transforming the soft callus into a bony callus of interwoven bone, forming a firm union within 2-3 months. This process is similar to the normal process of bone replacement found in all higher vertebrates but remodeling is accelerated in the vicinity of the injury.

PART 3 Loose and Dense Connective Tissue

p. 32-8 **a.** Yes. These cells contribute to the matrix of connective tissue, phagocytize foreign particles, and participate in immune responses.

PART 4 Adipose (Fat) Tissue

p. 32-9 **a.** The cells are empty. They contain no lipid material (stained red or purple) because normal tissue preparation uses alcohol and lipid solvents to dehydrate the tissues—this process removes neutral fat from the vacuoles. Cell membranes and nuclei can be seen.

PART 5 Blood

p. 32-11 **a.** At least two. Neutrophils and lymphocytes are the most obvious cells on slides stained with Wright's stain.

  b. In mammals, red blood cells contain no nuclei while white cells have nuclei of various shapes (ovoid to sausage-shaped).

  c. Biconcave discs.

  d. Yes. Yes. No.

  e. Human red blood cells are smaller, biconcave, and lack nuclei.

EXERCISE C   Muscle Tissue

PART 1 Skeletal Muscle

p. 32-12 **a.** The cross-striations represent the alignment of myofibrillar structures to form functional units (sarcomeres) within the fiber.

  b. The alignment of myofibrils and membrane elements (T-tubules, sarcoplasmic cisternae) facilitate the rapid activation of striated muscle organelles and their coordinated contraction.

PART 2 Cardiac Muscle

p. 32-13 **a.** The intercalated discs are the junctions between adjacent cells. Slow action potentials are propagated along cardiac muscle cells and across the cell-cell junctions without appreciable delay, linking all areas of the heart in a coordinated fashion to produce a functional contraction.

Table 32C-1 Muscle Tissue Characteristics

Characteristic	Skeletal	Cardiac	Smooth
Shape of muscle fiber	elongated fibers	elongated, branching fibers	spindle-shaped fibers
Number of nuclei per fiber	many	1	1
Location of nuclei	peripheral	central	central
Function of muscles	produce movement (behavior); voluntary muscles, under conscious nervous control	maintains rhythmic contractions of the heart, spontaneously active; involuntary control	viseral and vascular movements; "goose-bumps;" some are spontaneously active; involuntary control
Location	axial and appendicular areas, anterior and posterior sections of digestive system, etc.	heart and major blood vessels near the heart	viscera, blood vessels hair follicles, glands, etc.

PART 3 Smooth Muscle

p. 32-13   a.   Yes

        b.   Long cells arranged parallel to each other. (See notes in the *Preparator's Guide*— portions of this exercise beyond 2. pertain to skeletal muscle and should be included in Part 1).

        c.   Skeletal

EXERCISE D   Nervous Tissue

p. 32-15   a.   No, you cannot see an *entire* motor neuron.

        b.   The long fibers project beyond the plane of the slide (fibers are up to 1 m long in some motorneurons).

        c.   In your fingers and toes.

Laboratory Review Questions and Problems

1. Epithelial, connective, muscle, and nervous tissues.

2. Connective tissue is composed of the most diverse cell types (includes cartilage, bone, loose and dense connective tissue, adipose tissue and blood).

3. All connective tissues are derived from embryonic mesenchyme cells. They continue to be regenerated actively throughout life from mesenchymal precursors. Connective tissues are characterized by their intercellular matrix which ranges from loose fibers and ground substance (areolar connective tissue) to dense fibers (dense connective tissues) to the addition of chondroitin sulfate (cartilage) or calcium/phosphate salts (bone), to a continuous, fluid extracellular matrix (blood). In adipose tissues, neutral fat is contained within the fat cells which are distributed with other connective tissues (beneath the skin, in mesenteries, and in the buttocks and mammary glands of females).

4. Loose connective tissue is typically found between the muscles of the body and the dermis of the overlying skin as well as in mucous membranes and various visceral organs. Dense connective tissues forms tendons (attaching muscles to bones), ligaments (attaching bones to bones), and in the fibrous capsules of organs and joints.

5.

Types of connective tissue	Cell types in each tissue	Location in human body	Appearance of tissue	Major function
1. cartilage	chondrocytes	articular cartilages, junction of ribs and sternum	hyaline matrix (with fibers); cells in lacunae; no blood vessels/nerves	organize growth during development; form resilient junctions between bones
2. bone	osteocytes	entire bony endoskeleton	compact outer bone with Haversian systems, inner spongy bone; cells in lacunae connected by canaliculi; vascular	support; protection; in conjunction with muscle system, produce movement (behavior)
3. loose connective tissue	fibrocytes (various blood cells)	beneath skin, around various organs	loose fibers, scattered cells	bind organs of the body together; structure
4. dense connective tissue	fibrocytes	tendons, ligaments, dermis, organ capsules	dense collagen fibers (regular or irregular)	support; movement; protection
5. adipose tissue	adipocytes with large vacuole	fat (subcutaneous, mesenteries, etc.)	sparce cytoplasm and matrix; large vacuoles	store energy; cushion organs; retard heat loss
6. blood	red and white blood cells suspended in plasma	circulatory system (heart, vessels)	red fluid with variety of separate cells; clots when exposed to air	carry oxygen and carbon dioxide, nutrients, wastes, hormones to and from all parts of the body; buffer tissues

6. Esophagus: (nonkeratinized, wet) stratified squamous epithelium (the stratified squamous epithelium of the epidermis is dry and is highly deratinized)

   Lungs: simple squamous epithelium in the alveoli proper (ciliated pseudostratified epithelium lines most of the upper respiratory tract)

   Blood vessels: simple squamous epithelium (endothelium)

   Kidney tubules: simple cuboidal epithelium

   Intestine: simple columnar epithelium (with microvilli = brush border)

7. Dermal (= membrane) bone forms by direct ossification within membranes. Endochondral (= replacement) bone is first patterned in cartilage which is replaced by erosion and ossification within and around the early organ. Dermal bone contributes to the outer bones of the skull and a portion of the pectoral girdle (clavicle) in mammals; endochondral bone forms the remainder of the endoskeleton. In fishes, however, dermal elements forms (bony) dermal scales and the dermal skeleton is a major structural element.

8. Bones contain blood vessels and nerves so a break produces bleeding (with hematoma formation and inflammation) and it hurts!

9. Bone is continuously broken down by large, multinucleate cells, the osteoclasts. It is continuously reformed by chondroblasts and osteoblasts with new matrix replacing resorbed bone and subsequently ossifying with the deposition of calcium salts. Trapped osteoblasts become mature osteocytes within the living bone tissue. In addition to continuous replacement, bone is reinforced where stressed (and weakened where it is not stressed).

10. The matrix of cartilage contains resilient collagen fibers, ground substance, water (up to 80% of the tissue is water), and the gel-like chondroitin sulfate. Bone matrix includes collagen fibers, ground substance, water, and rigid crystals of calcium phosphate salts (= hydroxyapitite) that harden the tissue and resist deformation.

11.

Skeletal	striated	voluntary	multinucleate	unbranched	rectangular-shaped
Cardiac	striated	involuntary	uninucleate	branched	rectangular-shaped
Smooth	unstriated	involuntary	uninucleate	unbranched	spindle-shaped

12. Blood cells form from mesenchymal precursors and are suspended in a continuous extracellular matrix (plasma). The major functions of red blood cells include gas transport (oxygen and carbon dioxide). The major functions of white blood cells include phagocytosis of foreign organisms or protein and cellular debris, and recognition of foreign materials, production of antibodies, etc. as major components of the immune system.

13. A typical nerve cell (motoneuron) is shown as Figure 32D-1. No. There are a variety of types found in the peripheral and central nervous systems.

Notes:

Laboratory 33

Introduction to the Study of Anatomy, and the External Anatomy and Integument of Representative Vertebrates

EXERCISE A   Life History and External Anatomy of Four Representative Vertebrates

Shark

p. 33-5   a. Water would move out of the shark's body into the seawater. In many marine bony fishes, the blood is actually hypoösmotic. Water loss is retarded by protective dermal scales and the reduction (or loss) of kidney glomeruli (which filter blood to produce urine) coupled with the excretion of salt by special salt-secreting cells located on the ills.

p. 33-6   b. The frog, turtle and shark each have a cloaca.

c. The human urogenital system has separate, anatomically distinct openings to the outside for the digestive (anus) and urogenital systems (vestibule in females, penile urethra in males). The external anatomy of the reproductive system of human males (penis and scrotum containing testes) and females (clitoris and vestibule with the opening of the vagina) is sexually dimorphic.

p. 33-7   d. Electric fields produced by other organisms or modifications to the organism's own field induced by nearby objects give information about the environment that facilitates orientation and movement, particularly in turbid waters where vision is reduced. Certain fishes possess "electric organs"—modified muscles or neuromuscular junctions—that may also be used in communication (long trains of discharges, like bird songs, may help individuals locate mates, repel competitors, and convey information). In a few specialized fishes (such as the electric ray, catfish, and eel)., electric organs are able to produce sufficient current to aid in feeding by stunning nearby prey (and to provide protection).

e. Internal fertilization allows the female to retain the developing egg—to be viviparous. Viviparity improves the pup's chances of survival by protecting it until it is well developed and capable of finding food and escaping predators.

f. The male shark is identified by pelvic fins modified as "claspers." Female sharks have broad and unmodified pelvis fins.

e. Integument or skin.

Frog

p. 33-8   a. The frog, turtle and shark each have a cloaca.

b. The human urogenital system has separate, anatomically distinct openings to the outside for the digestive and urogenital systems. The external anatomy of the reproductive system of human males and females is different.

    c. The frog has small, homodont (vomerine) teeth arranged in two series, one at the margins of the jaw and a medial series on the palate and the lower jaw. Homodont teeth are not functionally differentiated as are the heterodont teeth of mammals.

    d. The frog has external and internal (opening into the mouth cavity) nares. The shark has only external nares. The significance of this difference is that the shark's nares, which contain sensory epithelium, are used only to mediate the sense of smell, whereas the frog's nares also provide a passageway for movement of air between the outside and the lungs.

p. 33-9   e. Frogs lack external genital structures. During breeding, the base of the thumb is enlarged and the forearm is swollen. Males may be smaller than females of the same age. Confirm the sex of your specimen in Laboratory 36 (p. 36-14).

    f. The lateral line system disappears at metamorphosis.

    g. The fore and hindlimbs differ in overall length and the length of individual major bones. The hindlimbs are relatively longer in all measures.

    h. The hindlimbs of the frog are modified for jumping and swimming.

    i. They are composed of jointed segments capable of independent movements (and contain skeletal elements homologous with those of other tetrapods).

    j. The moist, vascularized skin facilitates oxygen diffusion into the body and may take up water in the moist environments to which most amphibians are restricted.

Turtle

    a. Birds and mammals are endotherms, characterized by the relative constancy of their body temperature. Most birds incubate their eggs by applying a portion of the body surface (brood patch) directly to the eggs and transferring heat from their body; female mammals carry developing young within their body—thus, all young would be exposed to the same thermal environment and there would be no way to control the sex of the offspring and ensure that both male and female young are produced.

    b. In the painted turtle, the male's claws are longer than the female's and the plastron has a depression allowing him to fit on top of the female during mating. Males are usually smaller. Confirm the sex of your specimen in Laboratory 36 (p. 36-15).

p. 33-10   c. The frog, turtle and shark each have a cloaca.

    d. The human urogenital system has separate, anatomically distinct openings to the outside for the digestive and urogenital systems. The external anatomy of the reproductive system of human males and females is distinct.

    e. The turtle has external nostrils at the surface of the dermal bone and distinct nasal cavities opening into internal nares situated in a deep pocket behind the palate. The turtle lacks teeth.

    f. The forelimb has five digits. The hindlimb has five digits.

Rat

p. 33-11   a. Identify the female by the vaginal opening in front of the anus. Identify the male by the scrotal sacs located at the base of the tail and the penis in front of the anus.

    b. Partitioning the mouth cavity separates food from respiratory gases. The complete palate is also essential for nursing—young can produce reduced pressure that facilitates the flow of milk from the mother. The closed palate is thus characteristic of placental mammals.

p. 33-12 **c.** The rat's teeth are located at the margins of the jaw (mandible). They are surrounded by flaps of skin forming lips and cheeks (not found in the other representative vertebrates studied).

d. The shark has several series of homodont (similar) teeth, one series at the margins of the jaw and several series of medial teeth (located within the jaw). In contrast, the rat has only a single series of teeth at the margins of the jaw. Mammals have heterodont (different) teeth with different functions—incisors for cutting, canines for tearing, and pre-molars and molars for crushing. The numbers and structure of mammalian teeth vary in different groups and are correlated with diet.

e. Yes (they are said to be thecodont). No.

f. Four digits are on each forelimb. Four digits are on each hindlimb.

EXERCISE B   Integument—A Dynamic Interface

Shark

p. 33-16 **a.** The shark has placoid scales, which are similar to mammalian teeth in structure. Placoid scales (dermal denticles) are composed of a pulp cavity surrounded by dentine (composed of hydroxyapatite and 30% organic material) and coated with enamel (hydroxyapatite).

b. New cells originate in the basal layer of the epidermis.

c. Mesenchyme cells in the dermis produce the dentine ("cosmine") of scales. Epidermal cells secrete an outer enamel ("ganoine") layer on the developing scale. Although there are some structural differences between these layers in scales and teeth, they are generally regarded as being homologous.

d. Ganoid scales in which the outer layers are emphasized are found in several primitive ray-finned bony fishes (*Polypterus* and garpikes). The scales of bony fishes (cycloid or ctenoid scales) are thin discs of dermal bone (often largely reduced to fibers and matrix or osteoid) which develop in overlapping folds of the skin. Cosmoid scales in which the dentine-like materials are well developed characterize some of the ancestral lobe-finned fishes. The placoid scales of sharks are composed of dentine coated with "enamel" and generally lack dermal bone.

Frog

a. Keratin protein infiltrates epidermal cells as they age. More superficial dead cells are filled with keratin. Compared with the skin of amniotes, however, the epidermis of frogs is not highly keratinized (cornified).

p. 33-17 **b.** Secreted mucous protects the epidermis and reduces friction in water. The granular glands produce various secretions, which may be used for defense, as mating attractants, or (as in the Surinam toad) to nourish the young. Some amphibians have secretions that are highly toxic and may be aposematicaly (warningly) colored (e.g., "poison-arrow" frogs).

c. Amphibian diversity is greatest in warm, moist habitats, such as equatorial rain forests.

Turtle

a. The epidermis of the turtle appendages is extensively cornified. The skin of the appendages is covered with horny scales connected to each other by less heavily keratinized areas. These scales are modified on the turtle carapace and plastron to form large, horny scutes that cover fused plates of dermal bone.

b. The frog epidermis is not heavily keratinized, and is thus subject to rapid desiccation.

p. 33-18 **c.** The skin glands (mucous glands) in amphibians assisted in keeping the skin moist (though amphibians cannot live for long periods of time without water). Terrestrial vertebrates evolved mechanisms to reduce water loss from the skin. The development of an outer layer of dead cells filled with the water-insoluble protein, keratin, reduced epidermal evaporation. Glands are reduced to eliminate water loss as part of their secretions.

d. No. The boundaries of the surface horny scutes alternate with those of the underlying bony plates, as a means of strengthening the structure of the carapace.

e. The dermal scutes develop from dermal bone covered with keratinized epidermis. Fish scales consist of dermal bone, often reduced, and coated by varying amounts of dentine (cosmine) and enamel (ganoine).

Rat

p. 33-19 **a.** Sebaceous glands are usually associated with hair follicles (but may be present in hairless regions). They produce an oily secretion that lubricates and protects the hair and skin.

b. Human eccrine sweat glands produce a watery secretion that aids in evaporative cooling. These glands usually open directly to the surface of the epidermis. Apocrine sweat glands (located only in the axillary and pubic regions) are, like sebaceous glands, usually associated with hair follicles. Their secretions include apical portions of cell cytoplasm that is subjected to bacterial degradation to form various fatty acids that produce "body odor." Their products may include pheromones, chemicals that act as external chemical messages used to inform others of the identity, behavioral state, or reproductive condition of the individual. These types of "sweat" glands are variously distributed in other mammals—horses have eccrine glands and produce copious sweat that aids in cooling; dogs and rats have apocrine sweat (sudoriferous) glands and cool by panting (evaporative cooling).

c. The rat has 4 pairs of nipples. The number of nipples in mammals is correlated to the average number of young produced. Rats have more nipples than do humans because rats produce larger broods of young.

p. 33-20 **d.** Fat tissue is living and is vascularized. It is therefore a less efficient insulator than is dead epidermal products such as feathers or hair, which trap air close to the body and thereby reduce radiative cooling. In larger aquatic mammals such as whales and seals, hair is unable to retain an insulative shell of air so it becomes reduced and thick layers of fat ("blubber") and reduced peripheral circulation are used to retard heat loss from the body core.

SUMMARY

a. The phylogenetic trends in epidermal structures include: (1) a reduction of dermal scales; (2) a thickening of the epidermis and cornification of outer cells as an adaptation to life on land, and (3) a diversification in epidermal structures (hair, fur, feathers, scales) for a variety of functions including insulation in endotherms.

Laboratory Review Questions and Problems

1. Mouth: anterior, ventral, cranial.
   Toes of a pig (in relation to the limb): distal.
   Toes of a human (in relation to the limb): inferior, distal.
   Tail: caudal, posterior

2.

Organ Systems	Examples
Integument	skin, eccrine sweat gland, hair follicle
Sensory	eyes, ears, nose, lateral line organ, proprioceptor
Control: nervous system	brain, spinal ganglion, nerve
endocrine system	pancreatic islets, thyroid, pituitary, gonad
Digestive System	stomach, intestines, liver, pancreas
Respiratory System	gills, lungs
Circulatory System	heart, blood vessels, lymphatics
Urogenital System	kidneys, testes, ovaries

3.

	Chromatophores	Glands (examples)	Epidermal Structures
SHARK	Yes	Scattered mucous cells	Placoid scales
FROG	Yes	Granular and mucous glands	Naked skin
TURTLE	Yes	Few	Dermal scutes and horny scales
RAT	Yes	Apocrine sweat glands, mammary glands, sebaceous glands	Hair, claws

The skin of these animals is similar in the presence of chromatophores and (at least some) glands. The major differences are in the degree of keratinization and the types of epidermal structures (scutes, scales, hair).

4. The placoid "scales" (dermal denticles) of sharks are the evolutionary remnants of armored ancestors. They provide mechanical protection. The thin, moist, "naked" skin of frogs, bathed in mucus, aids in gas exchange (oxygen uptake, carbon dioxide loss) and retards water loss to some extent. The dermal scutes and underlying bony plates that are fused to form the plastron and carapace of the turtle provide mechanical structure and protection and reduce water loss. The keratinized epidermis reduces evaporative water loss in mammals. Hair provides sensory information, protection, and may be involved in retarding heat loss (thermoregulation).

5. The non-living outer layer of epidermis protects the inner, living cells in many ways. It can be abraded mechanically (and is replaced through wear or molt), it resists damage by a variety of chemicals, it retards water loss (or gain), it protects the body from invasion by microorganisms, it intercepts UV radiation, it may become involved in thermoregulation, it is shaped by underlying muscles and provides the contact of an organism with its substrate, it

may facilitate information exchange (communication), etc.

6. Powered flight demands a sustained metabolic power output to remain aloft. Muscles (and other tissues) function most efficiently at an optimal (and relatively high) body temperature so that a stable body temperature would be advantageous to flying organisms such as bats and birds (and pterodactyls?). The advantages of endothermy include a thermal environment in which enzyme optima may be tuned efficiently to maintain constant functioning and activity, the ability to live in environments too cold (or too hot) for ectotherms, etc. Disadvantages of endothermy include a high food intake (up to 10× that of ectotherms) just to maintain constant body temperatures. Advantages of ectothermy include lower energy requirements and an ability to reduce activities when conditions become unfavorable. Disadvantages include thermal limitations to a range between 0° (or higher) and 42 ± ° C. With increased body size, heat loss becomes a problem for both endotherms and ectotherms (heat loss is a surface phenomenon and increases only with the square of a linear dimension while heat production, related to volume, increase with the cube). Note that large ectotherms (like some living crocodilians, the monitor lizard and, possibly, extinct dinosaurs) may maintain a high and relatively constant body temperature without paying the energetic penalties of endothermy. With declining body size, endotherms (such as hummingbirds and shrews) reach a lower limit where body mass is insufficient to produce enough heat to exceed its loss over a relatively large surface. Ectotherms, however, can become much smaller (even microscopic) while maintaining normal functions.

Laboratory 34

The Anatomy of Representative Vertebrates: Behavioral Systems

EXERCISE A Sensors (Affectors)

PART 1 The Lateral-Line System

p. 34-4
a. The sensory cilia of the neuromasts respond to water movement caused by moving objects or organisms and thus supplement vision in making the shark aware of prey, predators, and obstacles in the environment. The anterior neuromasts also sense sounds (low-frequency vibrations—50-150 Hz) and electrical impulses in the water.

b. Electroreceptors also provide the shark with a cognizance of its body movements and thus aid in orientation and stability. Coupled with electric organs in some elasmobranch and teleost fishes, these organs provide a modality used for communication.

PART 2 The Inner Ear

p. 34-6
a. Calcium salts.

b. As the head moves, fluid within the canals stimulates hair cells and provides the organism with a perception of its movements. The sensors are tightly coupled to the head; inertia within the fluid provides a relative movement that indicates a change of position or acceleration.

c. Cochlear duct.

PART 3 Other Sensors of the Head

EXERCISE B Control Systems

PART 1 The Nervous System

p. 34-12
a. Reception of nerve impulses from olfactory sensors in the external nares.

b. In mammals, the cerebral hemispheres also receive sensory projections from the eyes and ear. They have direct motor connections to the spinal cord for initiating certain voluntary activities. They become the major "association centers" for behavior and complex thought.

c. The cerebral hemispheres of mammals are relatively larger and better developed, with more convoluted surfaces than are those of sharks.

d. The cerebellum is the largest part of the shark's brain. It functions in the most prominent activities of the shark, the coordination and regulation of locomotion, equilibrium, and body orientation.

p. 34-13
e. Somatic sensory, visceral sensory and visceral motor fibers travel through the dorsal root. Somatic motor (and some visceral motor) fibers travel through the

ventral root. Note that, in mammals, the dorsal root contains only sensory fibers and the ventral root contains both somatic and visceral motor fibers.

PART 2 The Endocrine System

EXERCISE C Effectors: Muscles and bones

p. 34-14 **a.** In terrestrial vertebrates, the pelvic girdle is tightly fused to the vertebral column (sacral vertebrae) to form a single functional bone articulating with the head of the femur (the pectoral girdle is only indirectly connected to the spinal column via the clavicle in some vertebrates and support depends on muscles and ligaments of the girdle to stabilize the shoulder joint). The structure of the long bones provides maximum support and may be repatterned in response to stress. The length of the functional segments (and the position of joints) are modified to keep the center of gravity positioned appropriately when standing, crouching, etc. Muscles form functional, often antagonistic, relations with the bones they move. Ligaments and tendons stabilize (synovial) joints. Etc.

PART 1 Muscles

p. 34-17 **b.** The acromiotrapezius draws the medial border of the scapula towards the vertebral column. An antagonistic muscle would be located on the medial surface of the scapula.

p. 34-18 **c.** Yes, yes.
d. Major blood vessels and nerves are located between muscles in the connective tissue fascia. Branches serve all tissues of the limb.
e. When dissecting muscles, natural separations occur between the white, translucent connective tissue fascia separating each muscle. When you are in a muscle, gray (red in fresh tissue) muscle fibers are exposed and separations become ragged and uneven.

p. 34-19 **f.** The trunk muscles are much reduced in mammals such as the rat.
g. Trunk muscles are the major effectors of locomotion in fishes such as the shark and are proportionally well-developed. Limb muscles provide support and locomotion for terrestrial tetrapods such as the rat. Therefore the limb muscles are well-developed in the rat and the trunk muscles are relatively less-developed.
. They become stabilizing structures that provide flexible support for the body axis (note that the axial skeleton becomes less mobile in the region of the thorax and may be immobile within the pelvic girdle.)

PART 2 Bones

p. 34-20 **a.** The skeletal arch in front of the spiracle has developed into the lower jaw which would interfere with its full development. The spiracle may have a small "pseudobranch" that further aerates blood traveling to the head. Spiracles are particularly well-developed in slow-moving sharks and are displaced dorsally in bottom-dwelling species such as the rays and some sharks where they may serve as an incurrent pathway when the mouth is otherwise occupied. Spiracles are smaller or lost completely in fast-moving sharks, which characteristically swim with mouths open to channel water across the gills (ram ventilation). The dorsal placement ensures that fresh water will flow into the pharynx to provide oxygen.

p. 34-21 **b.** Caudal fins transfer the sinuous movements produced by the axial muscles into work by pushing on the surrounding medium (water). The dorsal fins stabilize and regulate propulsion. Pectoral and pelvic fins function to control roll, pitch, and yaw (and the pelvic fins may be modified for sperm transfer in males).

c. Longitudinal fibers within each myomere contract, shortening the distance between myosepta and, through their coordinated interaction with other contracting fibers on the same side of the body and relaxing fibers on the opposite side of the body, produce a flexion toward the contracting side (the notochord is incompressible so it can only bend, not shorten). Contractions move down the body to produce S-shaped movements that move the caudal fin from side-to-side and transfer force to the surrounding medium to produce forward motion.

PART 3   Other Effectors

Laboratory Review Questions and Problems

1.

Sense Organ	Stimulus	Cranial Nerve	Brain Projection
Eyes	light	optic (II)	optic lobes (mesencephalon)
Ears	position, change in position	vestibulocochlear (VIII)	metencephalon (cerebellum)
Nose	smell	olfactory (I)	olfactory bulb (telencephalon)

2. The lagena in the shark is short whereas in humans the cochlear duct (homologous to the lagena) is extended and coiled. The shark has a simple sensor located within the lagena. The inner ear of the shark is sensitive only to low frequency vibrations and several neuromasts may be involved. In mammals, the lagena has become elongated and coiled. The basilar membrane contains numbers of hair cells along its length; these sensory cells are involved in transducing vibrations produced by higher frequency sounds into nerve impulses.

3. Sensory receptors (modified dendrites) in the skin perceive the stimulus. Heat is perceived by free nerve endings. Touch and pressure are perceived by Meissner's corpuscles. Receptors stimulate sensory neurons which relay impulses to interneurons of the central nervous system. Interneurons relay impulses to motor neurons which stimulate muscle cells in a muscle (an effector). The muscle contracts, moving the hand from the hot burner. This rapid response is known as a reflex arc—you are aware it happened but have no control over it.

4. Movement of vertebrates is accomplished by the contraction of muscles working against a skeletal system. Since muscles can actively contract, and passively elongate, the movement of a skeletal element in a limb is dependent on the contraction and relaxation of antagonistic muscles, which have opposite effects in altering the position of the bony element.

5. Vertebrate effectors include: muscles and skeletal elements working together, electric organs (modified muscle tissue) found in certain fishes; chromatophores (pigment cells) in ectotherms; and glands (photophores, also found in some fishes, are modified glands that produce light).

More:

Laboratory 35

The Anatomy of Representative Vertebrates: Digestive and Respiratory Systems

Shark

p. 35-4   a. The transverse septum.

Frog

     a. The parietal peritoneum is the thin, transparent, permeable epithelium ("serosa") lining the inside of the body cavity.

p. 35-5   b. Mesenteries are extensions of the peritoneum and contain small amounts of connective tissue, blood vessels and nerves. Mesenteries support the stomach, small intestine, liver, large intestine, spleen, pancreas, urinary bladder, ovary and testis and are continuous with the visceral peritoneum covering these organs. The kidneys (and adrenal glands) are outside the coelomic cavity and are said to be retroperitoneal.

     c. The pleuropericardial membrane and transverse septum. Yes.

Turtle

     a. The rigid body wall formed by the plastron is not flexible and could not be moved by muscles.

     b. The parietal peritoneum is the thin, transparent, permeable epithelium lining the inside of the body cavity.

     c. Mesenteries are extensions of the peritoneum and contain small amounts of connective tissue, blood vessels and nerves. Mesenteries support the stomach, small Intestine, liver, large intestine, spleen, pancreas, urinary bladder, ovary and testis.

Rat

p. 35-6   a. The parietal peritoneum is the thin, permeable epithelium lining the inside of the body cavity.

     b. Mesenteries are extensions of the peritoneum and contain small amounts of connective tissue, blood vessels and nerves. Fat may also be stored in the mesenteries of mammals.

     c. Mesenteries support the stomach, small intestine, liver, large intestine, spleen, pancreas, urinary bladder, ovary and testis.

Shark

p. 35-8   a. The oral cavity of the shark is much smaller than the pharynx. It terminates just in front of the openings of the spiracles.

p. 35-9   b. Stomach contents often include small fish.

p. 35-10   c. In other vertebrates the walls of the small intestine are lengthened and folded to increase the absorptive area. The surface is often correlated with diet, being relatively larger in vertebrates such as herbivores that feed on less digestible products (vegetation) than in vertebrates that feed on more digestible foods (carnivores, nectivores, etc.).

Frog

a. Just in front of the eustachian tube.

b. The oral cavity.

c. Bile is a fluid secreted by liver cells and stored in the gall bladder. Bile salts emulsify fats, contributing to their enzymatic digestion in the intestinal lumen. Bile also functions in excretion as liver cells extract waste organic compounds (such as the bile pigments, which are products of hemoglobin catabolism) and secrete them into the bile. The waste products move in the bile from the liver to the intestines and are excreted from the body through the digestive tract. Bile pigments are responsible for the dark color of feces.

d. Stomach contents may include insects and other small invertebrates.

p. 25-12   e. Dorsal mesenteries include the mesogaster, mesointestine, and mesocolon. Ventral mesenteries include the gastrohepatic ligament (lesser omentum) and falciform ligament. See Table 35A-1.

Turtle

a. The arrangement of the oral cavity and pharynx of the turtle is more like that of the frog. The arrangement in both the frog and the turtle allows for simultaneous ingestion and respiration. Both the frog and the turtle have oral cavities which are relatively larger than their pharyngeal cavities' paired internal and external nares which function in air passage between the lungs and the outside, and well-developed palates. These are primary palates forming the roof of the oral cavity and may be vaulted to provide some separation of food and respiratory gases. (Mammals have secondary palates that provide greater separation of naso- and oropharyngeal regions and allow infants to nurse—see **c.** below.)

b. The gall bladder is an expandable organ that stores bile secreted by the liver. Bile is released to the intestine from the gall bladder through the cystic duct and common bile duct.

Rat

p. 35-14   a. The salivary glands produce saliva, which contains mucus to lubricate food for ease in swallowing; and, in some vertebrates, salivary amylase which initiates starch hydrolysis.

b. The epiglottis prevents food from entering the trachea and lungs.

c. The digestive and respiratory systems are incompletely separated within the pharyngeal region of all vertebrates. This may be important in providing an alternate pathway for respiratory gases in case the nasal passages become blocked. The major disadvantage of the incomplete separation is the need to segregate ingested food from inspired air in terrestrial vertebrates. The epiglottis assists in this function by shunting swallowed food into the esophagus and away from the trachea. Lungs develop as pouches evaginating from the embryonic pharynx—thus, the confluence of these systems in vertebrates with lungs has a developmental origin.

p. 35-15   **d.** The anterior lining of the stomach is rough textured and contains striated muscle in the walls. The lining of the posterior portion of the stomach is smooth and glandular in appearance.

e. The striated muscle in the anterior portion of the stomach is under voluntary control and allows the animal to regurgitate unpalatable food.

p. 35-16   **f.** The omental bursa is a double-walled mesenteric pouch that develops from an overgrowth of the mesogaster to surround ventrally the major visceral organs in mammals. It offers added support and cushioning to the intestines. Yes, the omental bursa is vascularized and contains considerable quantities of fat. It often contributes to a human male's "middle-age spread."

EXERCISE C   The Respiratory System

Shark

p. 35-17

Frog

a. Both respire by means of gills. Water taken in through the mouths of tadpoles and most sharks and is passed over the gills.

b. Sharks usually have 5 pairs of gill slits plus a pair of modified pharyngeal slits, the spiracles, which may function to take in water which is then passed over the "internal gills" and out the external gill slits. Tadpoles have four pairs of gill slits and lack spiracles. In tadpoles, gill filaments project beyond the body as "external gills" and are covered by an operculum that serves to protect them and acts as a pump to assist in moving water across the respiratory surfaces.

p.35-19   **c.** The internal surface is smooth, highly vascularized (though not as extensively as mammalian lungs) and partitioned (septate). A trachea is present.

Turtle

a. The turtle lung is structurally more complex than is the frog lung. The air passageways in turtle lungs branch, with primary bronchi ramifying from the trachea into secondary bronchi. Frog lungs are septate but lack this extensive system of air passageways.

Rat

p. 35-20   **a.** The cartilaginous supports become smaller and incomplete in the finer diameter branches of the bronchi. The narrowest diameter bronchioles lack cartilaginous supports and are surrounded by smooth muscle tissue that aids in the distribution of air within the lungs.

b. The lung appears to be "spongy"—less solid than internal organs like the liver because of the extensive airways and alveoli (which contain air in the living rat).

Laboratory Review Questions and Problems

1. Homeostasis is the maintenance of a relatively stable internal physiology in face of changing internal and external environments. Homeostatic mechanisms include: maintenance of the oxygen content and chemical composition of the blood through controlled respiration; removal of waste products of digestion and metabolism; constancy of body temperature in endotherms; and osmoregulation of blood and body fluids.

2. Suspending mesenteries would constrain the movement of the heart and would thus mechanically inhibit the pumping of blood.

3. An increased surface area enhances absorption. Increasing the length of the intestine is one way to increase absorptive surface area. Increasing the complexity or folding of the intestine, exemplified by the spiral valve in the shark intestine, also increases absorptive surface area. Organisms having intestines that are relatively short in length characteristically have diets of microscopic or easily digested foods (such as nectar). Carnivorous animals have relatively longer intestines, since the increased surface area is needed for absorption of nutrients. Herbivorous animals have the longest intestines and the greatest absorptive surface areas.

4. The liver is an exocrine gland which functions in homeostasis to: (1) store and release carbohydrates and thereby regulate blood glucose levels; (2) manufacture plasma proteins to maintain the osmotic balance between blood and tissues; (3) produce bile, which assists in enzymatic digestion processes; (4) catabolize substances such as hemoglobin from dead red blood cells, hormones and foreign substances and then secrete the breakdown products into the bile which is excreted from the body through the digestive system and (5) deaminate amino acids to produce simple sugars or fatty acids.

5. The evolutionary trend in lung development, from the ancestral lungfishes to present day mammals is towards increased division of the lining of the lung from a simple, vascularized, air-holding sac (in primitive lungfishes) to the minutely compartmentalized alveoli with ramified system of bronchi in the mammalian lung. (The lung of birds has evolved in association with air sacs that ensure complete ventilation and that pass fresh air over the respiratory surfaces on both inspiration and expiration. Avian lungs themselves are relatively small and rigid—they need not inflate and deflate as the balloon-like lungs of other vertebrates do because they are attached to a pump (bellows-like) system that renews the air.)

6.

	Advantages	Disadvantages	Organisms
Gills	Moist respiratory surfaces, one-way flow	Water contains less oxygen than air	Fish; sharks; mammalian embryos
Lungs	Infolded, interior surfaces retard water loss; air contains more oxygen than water	Two-way flow of gases and consequent incomplete extraction of oxygen (birds are an exception	Amphibians; reptiles; mammals; birds

Notes:

Laboratory 36

The Anatomy of Representative Vertebrates: Circulatory and Urogenital Systems

p. 36-5   **Figure 36A-6** *Vertebrate circulatory systems.*

Fish (clockwise, from upper right): renal portal vein; renal vein; hepatic portal vein; hepatic vein; posterior cardinal vein; anterior cardinal vein.

Amphibian/Reptile (clockwise, from upper right): renal portal vein; ventral abdominal vein; renal vein; hepatic portal vein; posterior vena cava; hepatic vein; anterior vena cava.

Mammal (clockwise, from upper right): iliac vein; renal vein; hepatic portal vein; hepatic vein; posterior vena cava; anterior vena cava.

p. 36-7   **a.**   Arteries contain more smooth muscle tissue and elastin fibers than do veins. Arteries have thicker walls and smaller lumens than veins. Note that veins contain valves to keep blood flowing in the correct direction; arteries do not contain valves (flow depends on hydrostatic pressure maintained by the heart) but these features are not evident on the slide.

PART 3   Heart

Shark

p. 36-8   **a.**   The four chambers of the shark's heart (in order of blood flow) are the sinus venosus; atrium; ventricle, and conus arteriosus.

b.   The ventral aorta and afferent arterioles should be shaded.

c.   The human heart accommodates "parallel" respiratory (heart → lungs → heart) and systemic (heart → body → heart) circulation systems, in which oxygenated and less oxygenated blood are completely segregated. The human heart is composed of paired atria and ventricles. The shark's "serial" heart is composed of four chambers (answer **a.**) arranged after each other in a series. The sinus venosus contributes to the pacemaker (sinoatrial node) in humans. The semilunar valves, preventing backflow of blood from the aorta into the ventrcles, are derived from the conus arteriosus. Referring to the *number* of chambers is thus uninformative.

Frog

a.   Shade three chambers—two atria and the ventricle.

p. 36-9 **b.** Three (two atria, the ventricle) or five (sinus venosus, two atria, ventricle, and conus arteriosus). This answer doesn't tell you much.

c. The walls of the ventricle are thick; the atrial walls are relatively thin.

d. The ventricle is muscular in order to pump blood out of the heart and into the systemic circulation (long pathway, friction—resistance to flow). The atria serve as holding sacs for blood entering the ventricle and need only move blood into the adjacent ventricle—thus, they are not as muscular as the ventricles.

Turtle

a. The wall of the right atrium contains the sinoatrial node which initiates contractions.

Rat

p. 36-10 **a.** The heart valves are flaps of endothelium and connective tissue lining the lumens of vessels that open to allow blood flow and close to prevent backflow. Thus, valves are passive structures, although in some higher vertebrates valves may contain some muscle fibers.

EXERCISE B   Urogenital System

p. 36-10 **a.** Both embryonic and evolutionary development tends towards modification of the posterior kidney tubules to form the functional kidney elements, and (excluding bony fishes) modification of the anterior tubules to form ducts serving the male reproductive system.

Shark

p. 36-13 **b.** The uterine lining is folded and highly vascularized.

c. Infants are bound by a yolk stalk to their individual yolk sac (an extraembryonic structure that stores the bulk of the yolk available to the developing embryo). Rich vascular beds in the yolk sac and uterine lining are closely opposed to facilitate exchange of gases and wastes. There is, however, no direct connection of maternal and fetal tissues.

Frog

p. 36-14 **a.** Frog eggs are shed into the water and fertilized externally. Therefore, copulatory structures are lacking. (The male grasps the female, aiding in the laying process, and deposits sperm as the eggs emerge. This is called amplexus.)

b. Elements of male and female gonads and their duct systems appear in both sexes and develop through an "indifferent" stage. Later, as environmental conditions tip the balance or as appropriate genes become expressed, development in one direction or the other (male or female) predominates. However, this occurs relatively late in the overall maturation of the organism and it is common for vestiges of the other sexes' system to remain.

c. (While adrenals are whitish in fresh specimens, they are likely to look like the kidney in preserved specimens and are difficult to distinguish.) Adrenal glands consist of two distinct components, the internal chromaffin tissue (the adrenal medulla in mammals) which secretes the catecholamines epinephrine and norepinephrine, and the surrounding interrenal tissue (called the adrenal cortex in mammals) which secretes steriod hormones. These secretions affect: metabolism of minerals such as sodium and potassium; metabolism of carbohydrates and proteins; stress responses; and sex hormone activity. These two components appear as separate elements during phylogeny but become more closely associated to form the compound endocrine gland in mammals called the adrenal.

Laboratory Review Questions and Problems

1. The circulatory system functions in internal transport of blood to all body organs, in order to deliver nutrients and oxygen and remove wastes and carbon dioxide. Functional exchanges occur in thin-walled capillaries or sinusoids. The circulatory system also monitors various products (glucose, carbon dioxide, etc.) and participates in their homeostatic regulation. Blood carries hormones and other information-containing molecules. The circulatory and lymphatic system (which returns fluids from interstitial spaces) is also involved in clotting and inflammation, tissue repair, regeneration, defense against foreign materials or organisms, and immune responses.

2.

Arch	Shark	Frog	Turtle	Rat
I	Mandibular arch or spiracular artery (afferent portion absent)	disappears in early development	absent	absent
II	Hyoid arch	disappears in early development	absent	absent
III	Afferent and efferent arteries of the gills	internal carotid artery	internal carotid artery	internal carotid artery
IV	Afferent and efferent arteries of the gills	aortic arch	aortic arch	aortic arch
V	Afferent and efferent arteries of the gills	absent	absent	absent
VI	Afferent and efferent arteries of the gills	pulmocutaneous arch	pulmonary artery	pulmonary artery

3. In cyclostome fish, blood flows directly from the body organs through the posterior cardinal vein to the heart. In sharks, the cardinal veins are disrupted anterior to the kidney. Blood flows into a capillary network around the kidney tubules and then to the heart through the posterior cardinal vein.

In reptiles and some amphibians, the renal portal system is partially degenerate. Blood flows either through or past the kidneys and does not pass through capillaries.

In mammals, the renal portal system is absent. Blood from the posterior extremities flows directly into the posterior vena cava and to the heart.

Thus the posterior vena cava is an evolutionary composite of parts of the posterior cardinal, renal portal and hepatic veins (and several other embryonic channels that fuse during development).

4. Vertebrate hearts are shown in Figure 36A-7. The sinus venosus has been reduced to the sinoatrial node and the conus arteriosus to the semilunar valves. The atrium and ventricle split into two halves, providing separate pulmonary and systemic circulation in mammals. Note that these two streams did not separate completely until respiratory functions of the skin or cloaca were abandoned (in birds and mammals).

5.

Structure	Shark	Frog	Turtle	Rat
Kidney	ophisthonephric	ophisthonephric	metanephric	metanephric
Kidney duct	ophisthonephric	ophisthonephric	ureter	ureter
Accessory ducts	accessory urinary duct[1]			
Testis ducts	vasa efferentia, Leydig's gland[2]	rete testis, vasa efferentia	rete testis → efferent ductules → deferent (archinephric) duct	
Oviducts	oviducts	oviducts	oviducts	fallopian tubes
Method of fertilization	internal	external (amplexus)	internal	internal
Copulatory organ	pelvic fins (claspers)/cloaca	none	penis/cloaca	penis/vagina
Site of development of young	"uterus"	externally (in water)	external amniotic egg	uterus
Maternal-fetal connection	yolk sac	none	none	placenta and umbilical cord

6. **Arteries** (clockwise from upper right): carotid; aortic arch; subclavian; heart; renal; aorta; caudal mesenteric; iliac.

Veins (clockwise from upper right): hepatic; inferior vena cava; hepatic portal; renal; femoral; pulmonary; anterior vena cava.

Human Heart (clockwise from upper right): innominate (brachiocephalic); left common carotid; left subclavian artery; pulmonary artery; pulmonary vein; left atrium; left ventricle; dorsal aorta; inferior vena cava; right ventricle; right atrium; superior vena cava; aortic arch.

[1] The accessory urinary duct drains the more posterior portions of the kidney which produce the bulk of the urine. This leaves the original opisthonephric (archinephric) duct to carry sperm in male sharks.
[2] Leydig's gland consists of modified mesonephric tubules between those carrying sperm and urine and secretes a seminal fluid.

Laboratory 37

The Basics of Animal Form: Skin, Bones, and Muscles

EXERCISE A   The Language of the Body

p. 37-2   **a.** Cranial or superior.
            b. Dorsal.
            c. Caudal or posterior.
            d. Proximal, distal

p. 37-4   **e.** Superior.
            f. Superior, anterior

p. 37-5   Figure 37A-2 Body planes (from top): frontal section, sagittal section (both are longitudinal sections); cross section
            Figure 37A-3 Body landmarks (clockwise from top right): axillary, brachial, flank, groin, femoral, pubic, abdominal, thoracic, cervical.

EXERCISE B   Vertebrate Coverings: Skin, Scales, Feathers, and Hair

Fishes

p. 37-7   **a.** New epidermal cells originate from the basal layer of the epidermis, the stratum germinativum, which is composed of mitotically active cells.

p. 37-8   **b.** **Cosmoid plates** comprising the dermal armor or dermal "scales" of ancestral agnathan fishes were composed of dermal bone overlain by dentine or a dentine-like material produced by the dermis (cosmine) and enamel or an enamel-like material (ganoin) produced by the epidermis. With the loss of bone in cartilaginous fishes, the scales were reduced to small tooth-like structures of dentine with a thin layer of enamel surrounding a pulp cavity (like a tooth), forming **placoid scales.** Cyclostomes lost dermal scales altogether. Other fishes also reduced bone in the dermal scales, emphasizing either the dentine (cosmine) or the enamel (ganoin) components. *Polypterus* and the garpikes retain **ganoid scales** today. In teleosts, thin scales develop in overlapping skin folds. The scale appears to be composed of acellular bone underlain by dense fibers for flexibility. The surface of **cycloid scales** is patterned by a series of growth rings; **ctenoid scales** have comb-like projections on their posterior margins. The sculpturing of these scales may be formed by thin layers of dentine or enamel.
            c. Cycloid scales are a thin bony plates characterized by their round outline and circular growth rings ("cycles").

Amphibians

p. 37-8   **a.** No. The epidermis may have some keratin, but remains thin without special epidermal structures (claws are specialized epidermal structures composed of concentrated keratin)

    b. The mucous glands keep the skin moist, retarding water loss, providing protection, and also reducing friction in the water. Granular glands produce watery secretions, including irritants and poisons in some species. The secretions may be used for defense, as mating attractants, or to nourish the young (as in the Surinam toad). In amphibians (and other tetrapods), epidermal glands are multicellular with ducts opening onto the body surface; in fishes, most epidermal glands are unicellular.

Reptiles

p. 37-9   **a.** The epidermal scales are heavily keratinized plates while, between the scales, the skin is more lightly keratinized to facilitate independent movements of the scales.

    b. The frog epidermis contains little keratin and is thus subjected to rapid desiccation in dry air (but may pick up moisture in damp environments) while the skin of the snake is keratinized to prevent water flow in either direction. Some amphibians depend upon their moist skin for gas exchange; this is not possible in the keratinized skin of reptiles.

p. 37-10   **c.** No. The boundaries of the surface horny scutes alternate with the underlying bony plates as a means of strengthening the structure of the carapace.

Birds

p. 37-10   **a.** In down feathers, many barbs (with finer barbules) attach directly to the base of the feather retained in the feather follicle in the skin. These feathers lie beneath the contour feathers delineating the outer surface of the bird and provide insulation. In contour (flight) feathers, barbs attach to a long, keratnized axis extending from the base of the feather. In feathers of the wings and tail, these barbs have an interlocking mechanism consisting of hook-like keratin structures on barbules along one edge of each barb and fine, hair-like extensions on the opposing barbules. It is easy to separate barbs forming the vane of a wing feather by pulling them apart. A few strokes of the vane between your fingers or the mandibles of the bird relock the barbules to form a continuous vane. Most birds "preen" their feathers by manipulating them with their mandibles and transferring oily secretions from the uropygial or "preen" gland, a compound oil gland located at the base of the tail, to condition and waterproof the feathers. Note that feathers are homologous with reptilian scales.

Mammals

p. 37-10   **a.** Sebaceous glands are usually associated with hair follicles (but may be present in hairless regions). They secrete an oily product (sebum) that lubricates and protects the hair and skin.

    b. Human eccrine sweat glands open directly onto the surface of the skin and produce a watery secretion (99% water) that aids in evaporative cooling. Many mammals (including horses) have eccrine sweat glands that aid in heat loss. Humans also have a second type of sweat gland, usually associated with hair follicles and found in the axillary and pubic areas, the apocrine sweat glands. The products of these glands add some protein and lipid components to their product. Apocrine sweat gland secretions may function as pheromones, and coupled with bacterial action, produce body odor (lipids are degraded into long chain fatty acids). Dogs possess only this type of sweat gland and must rely on evaporative cooling from the tongue or wet body areas to maintain a constant body temperature.

p. 37-11   4.

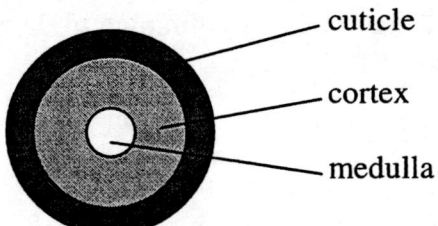

cuticle

cortex

medulla

This is a cross section of hair—whole mounts will show the imbricated scale-like surface of the cuticle covering the hair. While epidermal scales and feathers are regarded as homologous, hairs are believed to have evolved from tactile sensory structures associated with epidermal scales.

    c. Fat tissue is living and vascularized and blood carries some heat to the more superficial regions of the body while maintaining metabolism and gas exchange in the tissue. Neither hair (fur) nor feathers contain a vascular supply and can trap air, a most efficient insulator, over the surface of the body to retard both radiative and convective heat loss.

Summary

p. 37-12   **a.** Phylogenetic trends in structures of the integument include 1) the reduction and loss of dermal scales with the transition to land (dermal bone may persist in the skull and pectoral girdle); 2) a transition from unicellular gland cells lying in the outermost layer of the epidermis to multicellular glands removed from the surface; 3) a thickening of the epidermis and cornification (keratinization) of the superficial cells as an adaptation to life on land; and 4) a diversification in epidermal structures (epidermal scales, feathers, and hair).

EXERCISE C   Bones and Joints

PART 1   **Bones**

p. 37-12   .   This section assumes you will be using mammalian bones (other vertebrates do not have epiphyses but possess similar structures otherwise).

    a. Compact.

    b. No.

    c. Yes. Sharpey's fibers extend from the fibrous layer of the periosteum into the bony matrix to tightly anchor the sheath to the bone.

    d. Spongy bone.

    e. Cartilage. Endochondral bone preforms in cartilage. However, dermal or membrane bone ossified directly. This remnant of the dermal skeleton contributes bones that form the roof and sides of our skull and the clavicle (plus up to three other bones in the pectoral girdle of other vertebrates).

    f. Most cartilage is replaced by bone (it persists on the articular interface of long bones, in the junctions of ribs to the sternum, in the nasal cavity, in the larynx and trachea, and in intervertebral discs). As a living tissue, bone fractures can be repaired and tissues are continually repatterned by the removal and replacement of tissue.

p. 37-13   g.   This reduces the density of the bones to facilitate flight. Bird skeletons are said to be "pneumatic." Note that flightless and swimming birds may reduce the pneumaticity of their bones.

PART 2   Joints

p. 37-14

Location	Type of joint
Skull	immovable
Knee	synovial
Shoulder	synovial
Spine	slightly movable
Sternum	immovable
Wrist	synovial
Ankle	synovial
Hip	synovial

EXERCISE D   Muscles and Bones Working Together

PART 1   Isotonic and Isometric Muscle Contractions

p. 37-16   2.   HYPOTHESIS 1: shoulder and upper arm muscles involved in holding the arm outright will increase in size (diameter or thickness); antagonistic muscles will decrease in size (diameter or thickness) as they stretch
NULL HYPOTHESIS: muscles will not change in size
HYPOTHESIS 2: muscles of the upper arm involved in flexing the antebrachium (including the biceps) will thicken, opposing muscles (the triceps) will decrease in size or thickness
NULL HYPOTHESIS: muscles will not change in size
HYPOTHESIS 3: same as hypothesis 2
NULL HYPOTHESIS: muscles will not change in size
HYPOTHESIS 4: same as hypothesis 1
NULL HYPOTHESIS: muscles will not change in size

   a.   Isotonic to isometric; isotonic, isotonic, isotonic to isometric.
The independent variable is the type of movement.
The dependent variable is change in muscle size

p. 37-17 Table 37D-1

Movement	Left Arm	Right Arm	Isotonic or Isometric
Extend arms outward at shoulders and hold	29 cm	30 cm	isometric
Flex arms with palms upward	30 cm	31 cm	isotonic to isometric
Flex arms while grasping and 20 pound barbell	33 cm	34 cm	isotonic to isometric
Extend arms outward while holding a 10 pound barbell	30 cm	31 cm	isometric

    b. Yes, yes, yes, yes.

    c. Yes. The side of the body demonstrating "handedness" shows the greatest increase in circumference. This is because continued use of the part of the body has increased the muscle mass.

    4. Exercises: stand in the middle of a door frame and push on both sides; tense leg muscles; grasp hands and pull outward with arms (any exercise that develops muscle tension without movement).

PART 2 Muscle Fatigue

p. 37-17   **a.** The muscle fatigues more quickly.

    b. Yes, the fatigue effect is even more dramatic. As fatigue develops, the shortening and relaxing processes of the muscle become slower, especially the relaxation process. Because of this, the muscle fails to regain its original length before contracting again. The cause of the fatigue is the accumulation of waste products (CON, lactic acid, pyruvic acid, and acid phosphatase) as well as the depletion of energy-furnishing materials such as glucose.

    c. One hand fatigues faster. The less used hand has less conditioned musculature so it is unable to perform as well.

p. 37-18 Table 37D-2

	Minute 1 (in seconds)				Minute 2 (in seconds)				Minute 3 (in seconds			
	0-15	16-30	31-45	46-60	0-15	16-30	31-45	45-60	0-15	16-30	31-45	46-60
Right Hand												
Trial 1	24	23	24	22	22	22	21	21	21	21	21	20
Trial 2	26	24	22	20	21	20	20	20	16	17	16	16
Trial 3	30	24	21	20	20	20	19	15	16	15	15	15
Left Hand												
Trial 1	27	25	25	24	21	21	22	20	21	22	21	20
Trial 2	27	24	23	20	20	20	22	19	20	18	18	18
Trial 3	26	25	23	20	20	19	19	17	17	16	15	15

EXTENDING YOUR INVESTIGATION: Interactions of Muscles and Bones

p. 37-18   HYPOTHESIS:

        Muscle 1 The pronator superficialis muscle will rotate the lower arm forward.

        Muscle 2 The flexor digitorum superficialis will bend the wing tip toward the body.

Muscle 3 The extensor metacarpi ulnaris will straighten the wing tip out (extending it to line up with the middle portion of the wing).

Predictions: the hypothetical actions will be performed

p. 37-19   RESULTS:

Muscle 1 Hypothesis supported

Muscle 2 Hypothesis supported; bends in plane of wing

Muscle 3 Hypothesis supported

EXERCISE E   The Biochemistry of Muscle Contraction

p. 37-19   **a.** Many mitochondria—they are responsible for ATP production and muscle contraction is dependent upon the supply of ATP.

b. H zone will be reduced or disappear.

c. I band will decrease in size

d. I band decreases in size.

p. 37-21   **e.** Ten nuclei (student answers will vary with the preparation).

3. See Figure 37E-1(d) and answer 7 under Laboratory Review Questions and Problems.

8. HYPOTHESIS: The muscle elements will contract.
NULL HYPOTHESIS: The muscle elements will not contract.
Independent variable—the addition of ATP and Ca^{2+}.
Dependent variable—contraction of the muscle.

f. It remains the same.

g. No.

p. 37-22   **Table 37E-1**

	Myofiber Length		
	Before	**After**	**Change in Length**
ATP/Ca^{2+}	7 mm	7 mm	0 mm
KCl, MgCl$_2$	6.5 mm	6.5 mm	0 mm
ATP/Ca^{2+} and KCl, MgCl$_2$	8 mm	5 mm	3 mm

h. No, the muscle was expected to contract.

i. No, yes.

j. No; no.

12. HYPOTHESIS (KCl and MgCl$_2$ alone): There will be no response
NULL HYPOTHESIS: There will be no response
HYPOTHESIS (ATP, Ca^{2+}, KCl, and MgCl$_2$): The muscle strand will contract
NULL HYPOTHESIS: There will be no response
Independent variable—solutions used in the treatment.
Dependent variable—degree of muscle contraction.

14. The preparation contracted. See Figure 37E-1(e) or answer 7 in the review questions that follow.

p. 37-23   **k.** All solutions, ATP, Ca^{2+}, KC1, and MgCl^2 are necessary for myofiber contraction.

1. Anterior, distal, posterior (caudal), dorsal.

2. Characteristics:

	Chromatophores	Glands	Integument Structures
fishes	yes	scattered unicellular mucous glands	cosmoid scales, other dermal scales
amphibians	yes	mucous and granular glands	naked skin
reptiles	yes	few	epidermal scales
birds	yes	uropygial gland	feathers, scales, claws
mammals	yes	sweat glands, sebaceous glands, mammary glands	hair, claws, true horns

Similarities:
   all possess chromatophores; skin includes epidermis and dermis; skin is the major environmental interface, maintains the internal integrity, and protects the organism from physical, chemical, and biological elements in the environment

Differences:
   glands become multicellular and are reduced in terrestrial vertebrates; dermal structures become reduced as epidermal structures develop and elaborate; lateral line sensors, present in aquatic vertebrates, are lost with the transition to land

3. FISHES. The placoid "scales" (dermal denticles) of sharks are the evolutionary remnants of dermal scales found in armored ancestors. They serve as protection. The reduction of dermal bone reduces the density of chondrichthyes. Similar reductions in mass of the bony scales of osteichthyes protect the body while promoting flexibility. Some fishes have a mucous coat that actually streamlines the body in rapid swimming.

   AMPHIBIANS. The thin, naked skin of frogs aids in oxygen diffusion. Mucous glands help keep the skin moist and protect it and reduce friction in water. Granular glands may contain irritants or toxins that provide protection.

   REPTILES. The epidermal scales of reptiles provide protection. The dermal scutes and underlying bony plates of the plastron and carapace of the turtle reinforce the skin and offer added protection. The reduction of glands prevents water loss and a keratinized skin retards water loss through the skin and desiccation.

   BIRDS. Feathers of birds serve as insulation and retard heat loss. They form a surface supporting flight. The uropygial gland provides lubricant and conditioners for the feathers.

   MAMMALS. The keratinized epidermis of mammals reduces water loss. Hair aids in thermoregulation on land by retarding both heat and water loss. Horns may be involved in protection and mating behavior. Subcutaneous fat provides energy reserves. Eccrine sweat glands aid in evaporative heat loss. Sebaceous glands condition hair and the epidermis. Mammary glands provide milk for young.

   In all vertebrates, pattern and color (in those with color vision—many higher fishes, birds, higher primates) establish species and individual identities, may warn potential predators (aposematic colors), may foster crypsis, and may convey behavioral information.

4. The non-living outer zone of epidermis protects the inner layers of living cells.

5. The growth zone of the bone is in the epiphysis. It is composed of cartilage that is replaced by bone during growth. Severe trauma to this region can result in a discontinuation of the lengthening process. Epiphyseal (secondary) bone growth usually ends somewhere between 18 and 21 with the closure of the epiphyseal plates. Non-mammalian vertebrates have the potential to continue long bone growth throughout life because they have no epiphyses.

6. Exercises that do not require a great deal of movement—very different from isotonic and aerobic exercises that produce increased "fitness" (improved circulation, richer supplies of glycogen and myoglobin, etc.). People with back injuries or other skeletal problems including arthritis might find this form of exercise an acceptable alternative. Isometric exercises play an important role in "body-building" and do lead to increased muscle mass (hypertryophy).

7.   Relaxed   Contracted

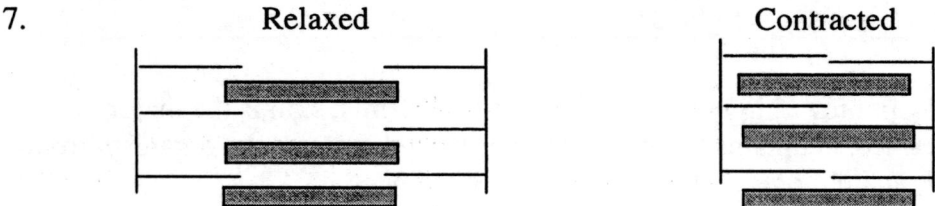

8. Hydrolysis of ATP by myosin molecules provides the energy for contraction and the "ratcheting" or sliding of myosin heads over the actin filaments. In order to split ATP $\rightarrow$ ADP + P$_i$, Mg^{2+} is needed by the myosin. During a contraction, Ca^{2+} (released from the sarcoplasmic reticulum) binds to troponin molecules, part of the thin myofilaments. Activated troponin moves tropomyosin and exposes the binding sites for myosin heads. With the exposure of binding sites on actin, the myosin heads bind to the actin and ratchet along, contracting the muscle ("sliding-filament" theory). Binding of a new ATP molecule to the myosin releases the myosin heads from their binding sites on the actin and the hydrolysis of ATP transfers energy to myosin for a new cycle of binding and contraction.

Laboratory 38

The Physiology of Circulation

EXERCISE A — Microscopic Examination of Human Blood Cells

p. 38-3
    a. The red blood cells (erythrocytes) are stained pink. No.

    b. Hemoglobin. Hemoglobin carries oxygen.

    c. 1-2 leukocytes/100× microscopic field. Blood from a healthy human usually contains 1-2 white blood cells for every 1000 red blood cells.

    d. Pink (orange in eosinophils).

    e. Lymphocytes.

    f. Eosinophils and neutrophils.

    g. Some of the red blood cells are ovoid or sickle-shaped.

    h. White blood cells are common and tend to lack dark-staining central areas.

    i. There are relatively more white blood cells (1~12/100 × microscopic field) in the smear of the blood from a person with mononucleosis.

    j. Monocytes are the most common.

    k. Frog red blood cells are larger, ovoid, and nucleated.

EXERCISE B — Pumps—The Vertebrate Heart

p. 38-6
    a. Arteries move blood away from the heart under relatively high pressure. The pulmonary artery, although carrying deoxygenated blood, is carrying the blood away from the heart to the lungs.

    b. The fetus's lungs are not inflated so they do not oxygenate the blood. Fetal blood is oxygenated in the placenta.

p. 38-7
    c. No. Arteries and veins are defined by the direction in which they carry blood.

    d. Arteries carry blood from the heart to other organs while veins return blood to the heart.

p. 38-8
    e. The left ventricle is responsible for pumping blood into the systemic aorta serving the body organs. It must be a powerful pump to overcome the added distance traversed and resistance in peripheral vessels.

    f. In the serial circulatory system of fishes, blood passes from the heart over the respiratory surfaces of the gills and then to the body. Friction in the vessels of the gills reduces the blood pressure available to move blood through the remainder of the body. In the parallel circulatory system found in birds and mammals, higher pressures can be maintained in both circuits to support greater needs for oxygen with higher activity levels and the energetic requirements imposed by endothermy.

EXERCISE C — Studying the Structure of Arteries and Veins

p. 38-9
    a. Arteries have thicker, stronger, and more elastic walls than do veins because the blood pressure within arteries (as the blood leaves the heart) is much higher. The

walls of veins are thinner to offer minimum resistance to blood as it flows back towards the heart.

    b. Valves help to prevent backflow of blood as it moves towards the heart.

    c. An atherosclerotic artery has walls that are rougher in texture and unevenly thickened. The lumen of the artery is partially occluded by plaque composed of lipids and fibrous connective tissue.

    d. Atherosclerosis interferes with the normal functioning of an artery by diminishing the elasticity of the artery and diminishing the diameter of the lumen, thus impeding the normal volume of blood flow from the heart to the body tissues.

EXERCISE D   Movement of Blood in a Goldfish Tail

p. 38-10   **a.** Blood moves more slowly in the capillaries.

    b. The diameter of a capillary is not much larger than the diameter of a red blood cell. The ovoid, nucleated red blood cells of a goldfish are approximately 12-14 μm in length and 8.5-9.5 μm in width. (In humans, red blood cells have diameters of 7-8 μm and thus must fold or twist to fit through the smallest capillaries which have diameters of approximately 5 μm.)

    c. As blood moves through the capillaries, gases (CO_2 and O_2) and other materials are exchanged by diffusion with the body tissues. The slow movement of blood through the capillaries facilitates this exchange.

EXERCISE E   Determining Blood Pressure

p. 38-11   **a.** Normal blood pressure is 120/80, but varies in individuals. Varied (individual) results will be obtained.

    b. Hyperventilation (chronic high blood pressure) may be a sign of plaque formation in the arteries (atherosclerosis).

    c. Hypertension is dangerous if allowed to continue indefinitely because it damages the endothelium of the blood vessels and thereby encourages plaque formation. Plaque can lead to narrowing of the lumen of arteries and increases the likelihood that a dislodged clot or plaque (an embolus) will cause a heart attack. Hypertension also damages the kidneys and ventricles of the heart and can cause a stroke by rupturing the walls of arteries and arterioles.

    d. Hypotension.

    e. Blood pressure in the thigh should be lower since the arteries of the thigh are farther away from the heart.

    f. Narrowing of the arteries increases blood pressure.

EXERCISE F   Measuring Pulse Rate and Blood Pressure

PART 1. Relation of the Heartbeat to Circulation

p. 38-12   **a.** Heartbeat rate equals the pulse rate.

    b.-c. This relationship exists because a pulse is the alternating expansion and recoiling of the elastic artery wall due to pumping of blood from the heart.

    d. A pulse can be found in any area of the body where a large artery is close to the surface of the skin.

PART 2. Variability of the Heart Rate and Blood Pressure

Table 38F-1 Pulse Rate (beats/minute)

Reclining	60
Immediately on standing	80
Standing	74
Sitting	68

p.38-13

a. The fall in blood pressure immediately upon standing results from the "falling" of blood from the upper body. To compensate for the gravitational effect, heart rate and blood pressure increase (thus blood pressure after standing for a period was greater than blood pressure immediately upon standing). Valves in the veins assist in preventing backflow of blood. However, the return of blood to the heart is aided by the constriction of veins between expanding muscles during exercise. Thus, the flow of blood back to the heart (and thus the general circulation) is slowed when standing and will increase immediately upon standing as the leg muscles "squeeze" leg veins. A more physically fit individual would exhibit a smaller difference in pulse rate upon standing. The stroke volume (the volume of blood pumped with each ventricular contraction) of a physically fit individual is greater.

b. The pulse rate of a sitting person is between that of reclining and standing since blood still must return from the legs, but the vertical distance is not as great.

PART 3. The Effect of Exercise on Heart Rate

	Before Exercise	Moderate Exercise	Heavy Exercise
Pulse rate	68	85	118
Rate of heartbeat	68	85	118
Time required for the return of sitting pulse rate		2-4 minutes	2-4 minutes

p. 38-14

a. It appears to be normal.

b. Strenuous exercise increases heart rate and raises blood pressure and could induce a heart attack or stroke.

PART 4. The Effect of Smoking on Heart Rate (*Optional*)

p. 38-14 **a.** Smoking increases heart rate.

EXERCISE G   Factors Influencing Heart Rate in the Water Flea, *Daphnia*

PART 1. The Effect of Temperature Change

p. 38-15 *Daphnia* **heart rate** (student data)

Temperature (° C)	Heart Rate (beats/minute)
5.5	108
8.0	120
11.5	128
13.5	192
16.0	228

sample calculation of Q_{10}

$$\frac{228}{108} \times \frac{10}{16-55} = 2.01$$

a. The Q_{10} for *Daphnia* at low temperature was 2. The Q_{10} for *Daphnia* at moderate temperature was 1.8. The Q_{10} for *Daphnia* at high temperature was less than 1.8.

PART 2. The effect of chemicals on Heart Rate

A Influence of Acetylcholine (A Neurotransmitter)

p. 38-15   HYPOTHESIS: Acetylcholine will slow the heart rate of *Daphnia*.
NULL HYPOTHESIS: Acethycholine will not affect the heart rate of *Daphnia*.

a. Acetylcholine will slow the heart rate of *Daphnia*.
Independent variable—effects of acetylcholine on heart rate

p. 38-16   Dependent variable—heart rate of *Daphnia*

Table 38G-1 Heart Rate

Time	Heart Rate
0 min.	200
1 min.	177
2 min.	170
3 min.	160
4 min.	149
5 min.	140

b. Yes. No.

c. Acetylcholine is a neurotransmitter which slows the heart rate in *Daphnia*.

d. It acts to restore the heart to its normal rate following emergencies.

B Influence of Epinephrine (A Neurotransmitter and Hormone)

p. 38-16   HYPOTHESIS: Epinephrine will accelerate the heart rate in *Daphnia*.
NULL HYPOTHESIS: Epinephrine will not affect the heart rate in *Daphnia*.

e. Epinehprine will accelerate the heart rate in *Daphnia*.

p. 38-17   Independent variable—effect of epinephrine on heart rate
Dependent variable—heart rate of *Daphnia*

f. Yes. No.

g. Epinephrine is a neurotransmitter which accelerates the heart rate in *Daphnia*.

h. Epinephrine would be produced in times of stress. Accelerated heart rate would assist the organism in active "fight-or-flight" responses to escape or avoid stress.

p. 38-17   ### Table 38G-2 Heart Rate

Time	Heart Rate
0 min.	190
1 min.	212
2 min.	288
3 min.	340
4 min.	360
5 min.	400

1. Left diagram (clockwise from upper right): artery, body cells, blood sinus, vein, heart.
    Circulation is open.
    Right diagram (clockwise from upper right): artery, body cells, capillaries, vein, heart.
    Circulation is closed.

2. Red blood cells are enucleate, smaller than most white blood cells, and contain hemoglobin.

3.

	Connective Tissue	Elastic Fibers, Smooth Muscle	Endothelium
Artery	moderate	thick	thin
Vein	moderate	moderate	thin

4. The peaks represent the systolic pressure. The troughs are diastolic pressure.

5. During the contraction of the ventricle, blood is forced into the proximal arteries under maximum pressure. Much of the energy is stored in the expanded elastic walls of these arteries and is measured as an increased pressure (systole). When the ventricle has completed its contraction, the valves separating the heart from the proximal arteries close, preventing back flow. The dilated walls of the arteries rebound elastically, releasing their stored energy and helping move blood into the arteries. However, the pressure in the vessels is relatively low at this time (diastole).

6. With a Q_{10} of 2, heart rate would double for every 10° C increase in temperature and halve for every 10° C decrease in temperature. There is no "normal" heart rate in an ectotherm, but there may be an optimal (maximal) rate near the upper lethal temperature. Warming would increase heart rate to the maximum tolerable body temperature; cooling would lower it to the lower lethal temperature (or to the freezing point of water).

7.

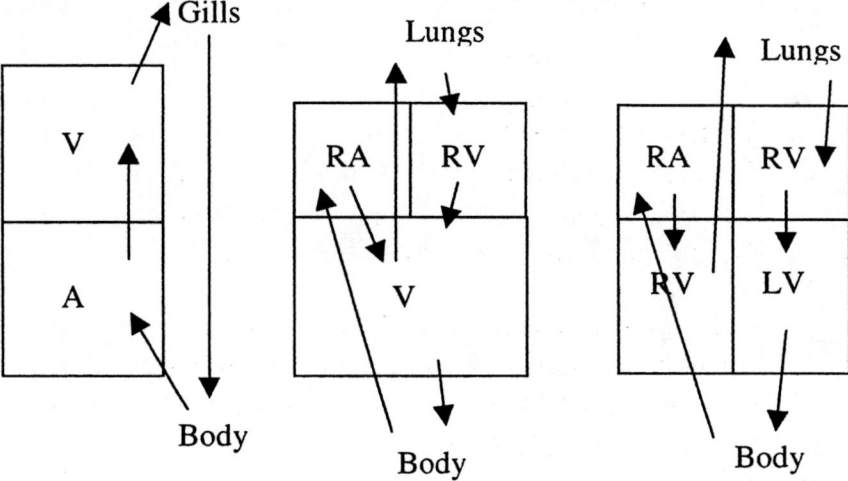

Notes:

Laboratory 39

Gas Exchange and Respiratory Systems

EXERCISE A   The Vertebrate Respiratory System

p. 39-2
a. The surface area of lungs increases, forming many more and smaller pockets.
b. Oxygen is exchanged across a moist respiratory surface. By increasing the surface area for respiration, more oxygen could be transferred from the air to the blood.

PART 1   How a Fish Breathes—Gills as a Respiratory Surface

p. 39-3
c. Lamellae increase the surface area for oxygen exchange.
d. Fish open and close their mouth as part of the respiratory cycle ventilating the gill lamellae with water containing dissolved oxygen. Muscles in the floor of the mouth and pharynx and the operculum assist in this process.
e. No. See Figure 39A-3.
f. If water and blood flow in the same direction, exchange of oxygen can only occur as long as there is a concentration difference between the amount of oxygen in the water and the amount of oxygen in the blood. The point at which there is NO concentration difference is reached quickly when blood and water flow in the same direction, but a difference, even a small one, always exists if blood and water are moving in opposite directions by counter-current flow. Thus more and more oxygen can be added to the blood as water flows over the gill lamellae.
g. The blood vessels in the gills are adherent and efferent arteries because they are carrying blood from the heart to the body organs. It is not until the blood returns from the capillaries of the body organs to the heart that we describe vessels as veins.

PART 2   How a Mammal Breathes—Lungs as a Respiratory Surface

p. 39-7
a. Cartilages get thinner and are not continuous. They form "C's" rather than encircling the air passages completely.
b. To keep the air passages open. Otherwise, the passageways would collapse under negative pressure during inhalation.
c. Number of lobes in the right lung—4
Number of lobes in the left lung—3
d. The tips of the lungs expand first, followed by the remainder of the lungs toward the trachea.
e. The pulmonary artery carries deoxygenated blood from the right ventricle.
f. Other arteries carry oxygenated blood from the left ventricle.
g. The pulmonary artery is an artery because it carries blood under pressure from the heart to peripheral organs (the lungs). Its thick walls and elastic tissue components are typical of all arteries.

p. 39-8
h. Elastic fibers are resilient but allow tissues to stretch under tension and return to their original conformation when tension is removed. The alveoli must be able to

expand and contract in order for the lungs to inflate and deflate as a part of breathing. Expansion also stores some of the energy expended on inhalation and yields it back to aid in expiration.

   i. Monocytes fight infection and ingest foreign particulate matter. Often, in polluted environments, foreign matter enters the lungs. This can also happen when one cuts the grass, sands wood, smokes a cigarette, etc. If the cilia do not move foreign materials up and out with the mucous that is coughed up, it is left to the macrophages to engulf and destroy the foreign material.

EXERCISE B   Respiratory Pigments

p. 39-8   **a.** An increase in hydrogen ion concentration $[H^+]$ decreases the pH of a solution (since pH $= -$ log $[H^+]$). The blood is buffered. Buffers are weak acids of bases (combinations of H^+ donors and H^+ acceptors) which function to maintain pH (by combining with free H^+ or OH^- ions in solution) when small quantities of acids or bases are added. Bicarbonate and carbonic acid act as base and acid to maintain blood pH.

p. 39-9   **b.** The red blood cells will hemolyze (rupture). Hemoglobin will be released into the solution. Centrifugation will separate the solution containing hemoglobin from the cell fragments.

   7. The initial color of the blood + yeast suspension is a milky red. Over the 5-10 minute incubation period, the color changes to purple-red.

   8. When the mixture is poured into a Petri dish, the color returns to red. The change in color is due to the oxygenation of the hemoglobin and the corresponding shift in the absorbance maximum (see graph in answer g).

   c. Yeast respiration (CO_2 production) simulates the condition of oxygen-depleted body tissues. Exposure to atmospheric O_2 allows hemoglobin to become oxygenated and CO_2 to diffuse from the solution.

   d. The PO_2 of blood plasma is higher than that of body tissues and oxygen diffuses from the blood plasma to the tissues.

   e. As CO_2 accumulates in the solution, the pH of the blood and yeast mixture decreases.

   f. As pH decreases, the affinity of the hemoglobin molecule for oxygen decreases. Hemoglobin in red blood cells readily gives up oxygen to the surrounding tissues as the pH of the blood plasma decreases (thus favoring the release of oxygen to supply the most deoxygenated tissues).

   g. The color change visible when deoxygenated blood is oxygenated is due to a shift in the abundances of the structurally different molecules, deoxyhemoglobin and oxyhemoglobin. The blue color of deoxyhemoglobin versus the red color of oxyhemoglobin would produce different absorption spectra (depicted on the next page).

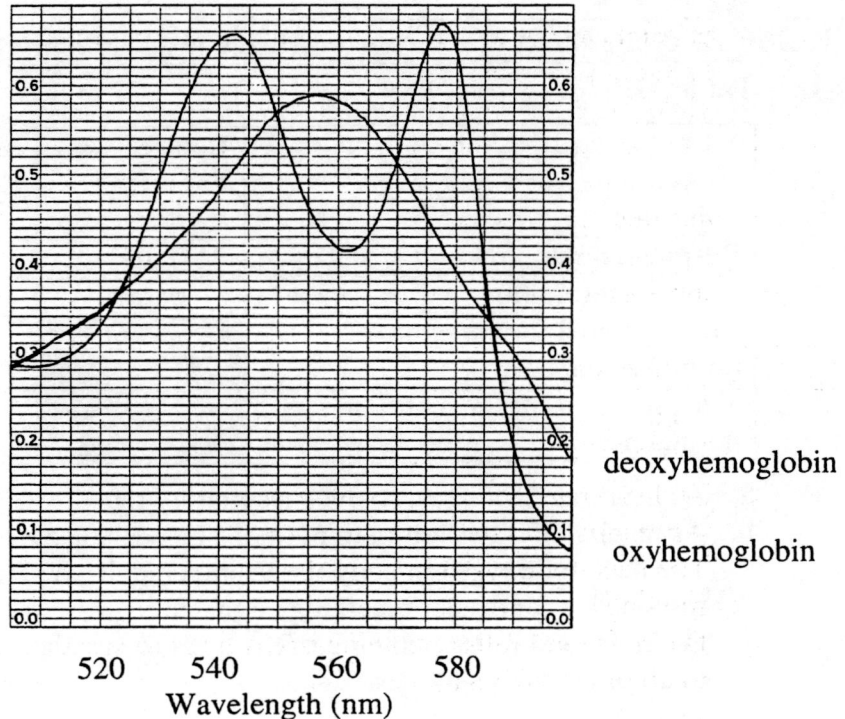

deoxyhemoglobin

oxyhemoglobin

520   540   560   580

Wavelength (nm)

EXERCISE C   Lung Capacity

p. 39-10   3. TV = 1.5 liters
               4. IRV = 2.2 liters

p. 39-11   5. ERV = 0.8 liters
               6. Vital Capacity = 4.5 liters
               a. This measured vital capacity of 4,500 ml agrees fairly closely with the average value for a 20 year old male (height 68 inches) which is 4,940 ml

---

EXERCISE D   How Does Smoking Affect Lung Capacity?

p. 39-11   HYPOTHESIS: Smoking will decrease the vital capacity of the lung.

p. 39-12   NULL HYPOTHESIS: Smoking will not decrease the vital capacity of the lung.
               Prediction—a smoker will have a reduced vital capacity.
               Independent variable—smoking.
               Dependent variable—vital capacity
               Results— IV = 1.8 liters
                         IRV=2.0 liters
                         ERV = 0.3 liters
               Vital Capacity = 4.1 liters
               Results support hypothesis but not the null hypothesis.
               Conclusion—smoking decreases lung capacity in a 20 year old male (height 72 inches) from the expected of 5.5 liters. Both inspiratory reserve volume and expiratory reserve volume are greatly reduced.

p. 39-13   Table 39E-1

	Before Exercise	Moderate Exercise	Heavy Exercise
Pulse rate	68	85	118
Respiration rate	16	18	24
Depth of respiration	normal	deeper	deeper
Time required for the return of normal respiration		2-5 min.	30 min.
Time required for the return of normal pulse		3 min.	4 min

a. As heart rate increases, so does respiratory rate.
b. As respiratory rates return to normal, so does pulse rate.
c. The need for oxygen in body tissues that require ATP during exercise is responsible for deeper breathing (signaled by respiratory centers in the medulla of the brain) and faster pumping of the heart to circulate the newly oxygenated blood to all of the body's tissues.

Laboratory Review Questions and Problems

1. The evolutionary trend in lung development, from the ancestral bony fishes to present day mammals, is towards increased division of the lining of the lung from a simple, vascularized air holding sac (in primitive fishes) to the minutely compartmentalized alveoli with a ramified system of bronchi in the mammalian lung.

2. This countercurrent flow enables the blood to pick up the maximum amount of oxygen possible from the surrounding water.

3. The oxygen concentration in lung alveoli is the same as the oxygen concentration in the pulmonary vein ($P_{O2} = 100$). The P_{O2} in the pulmonary artery bringing blood to the lung is much lower ($P_{O2} = 38$) so oxygen diffuses from the lung alveoli into the lung capillaries.

4. Hemoglobin (Hb) is composed of a globin molecule, composed of four polypeptide chains. Each chain is complexed with a heme group containing iron. The four polypeptides cooperate in binding oxygen. After the first O_2 molecule binds to the iron in one heme group, the Hb molecule changes shape, facilitating the uptake of additional O_2 until it is saturated, carrying four O_2 molecules (oxyhemoglobin). Unloading of one O_2 molecule facilitates further unloading (deoxyhemoglobin). Increased CO_2 (higher pH) also facilitates unloading of O_2 in tissues (the Bohr effect).

5. Iron is necessary for hemoglobin formation. Hemoglobin is the carrier of oxygen in red blood cells. If hemoglobin cannot be formed in proper amounts, fewer red blood cells are made and less oxygen is carried by the blood.

6. Carbon dioxide lowers the pH of the blood. Remember that
$$CO_2 + H_2O \leftrightarrow H_2CO_3 \leftrightarrow H^+ + HCO_3^- \leftrightarrow 2 H^+ + CO_3^{2-}$$
so that increasing amounts of CO_2 increase the $[H^+]$ and decrease pH.

7. H_2CO_3, carbonic acid, is a weak acid that can donate or accept H^+ and serves as a buffer in the blood (see equation in question 6). CO_2 is added in the tissues, shifting the equation toward the right. CO_2 is removed in the lung, shifting it toward the left. As P_{CO2} and pH increase, hemoglobin's affinity for O_2 is reduced, facilitating unloading of oxygen in the tissues.

8. The blood carries less oxygen than the amount in the alveoli. The P_{O2} will be slightly lower in the blood because the respiratory surface that allows for oxygen exchange has been damaged. Continued smoking leads to enlargement of the alveoli and deterioration of the alveolar walls with chronic inflammation and fibrosis that may lead to emphysema.

9. Pulse rate and respiratory rate increase with exercise. Increased ventilation of the respiratory surface and more rapid circulation combine to deliver needed amounts of O_2 to tissues and remove waste CO_2, lactic acid, etc.

10. As the diaphragm contracts, it flattens and increases the height of the thoracic cavity. Contraction of the external intercostal muscles lifts the rib cage and pulls the sternum forward, further increasing the volume of the thoracic cavity. The lungs are tightly coupled to the parietal peritoneum lining the pleural cavity so they also expand, reducing the gas pressure within the lungs below that of the surrounding atmosphere so fresh air flows into the lungs. It is the "negative pressure" within the lungs that produces this flow.

11. The vital capacity of the lungs is the total amount of air that can be exchanged with each ventilatory cycle. It is measured by an instrument called a spirometer. It is essentially a bell inverted over water that is displaced as you breathe into a mouth piece. Displacement is recorded on a rotating drum.

Notes:

Laboratory 40

The Digestive, Excretory, and Reproductive Systems

EXERCISE A   Examining the Digestive System

PART 1   Microscopic Anatomy of the Digestive System

p. 40-3   1. Yes. The cells are large; some are clear and others are filled with granular cytoplasm. See Figure 40A-2a for a photograph of salivary gland tissue.

    a. Starch (crackers—as amylase begins to act, their taste becomes "sweet").

    b. Glucagon promotes the breakdown of glycogen to glucose in the liver and increases blood glucose levels. Insulin stimulates the uptake of glucose by cells, thus decreasing blood glucose levels.

p. 40-4   3. Yes. Yes. In humans, the parenchymal cells of the liver form single layers separated by blood flowing from the renal portal vein (and hepatic artery) toward the central hepatic vein. The bile canaliculi are formed by the plasma membrane of adjacent parenchymal cells within the single-celled sheets and drain into interlobular bile ducts. Note the counter-current flow of bile and blood.

p. 40-5   **c.** The liver is the largest internal organ of the body. It stores and releases carbohydrates, processes amino acids, synthesizes essential proteins (such as enzymes and clotting factors), and is the major source of lipoproteins (HDL and LDL) to transport cholesterol and fats. It also manufactures plasma.proteins important to maintaining the osmotic properties of interstitial fluids. The liver stores fat soluble vitamins and produces bile salts to emulsify fats during the digestive process. It breaks down hemoglobin from damaged red blood cells to form bilirubin. It helps in detoxification of foreign substances and is important in hormone regulation. Because of its widely divergent physiological functions, few systems of the body remain untouched by the actions of the liver. For this reason, damage to the liver is likely to cause a variety of physiological problems.

    d. Cylindrical or elongated from the basal to apical (free) surface.

    e. Mucus secretion aids in movement of materials through the digestive tract by reducing friction and lubricating/protecting delicate tissues that line the lumen.

p. 40-6   **f.** A high fiber diet would require more enzymatic activity for digestion of fibrous materials. This, and the subsequent process of nutrient uptake, would take a longer time for completion. A longer intestine with a greater internal surface area facilitates this process and allows for complete uptake of nutrients from the digestive tract. (Note that vertebrates are unable to digest cellulose, however long the intestine. However, more advanced artiodactyls like cows have a specialized stomach that houses a "soup" of microbes that can and the resulting products, including short-chain fatty acids, can be absorbed on their subsequent passage through the gut. Cows belong to a group of "foregut" cellulose utilizers. Cows do

---

require some green forage in their diet. Horses, on the other hand, have a large cecum and large intestine that houses various symbionts that attack cellulose. While this is less efficient, horses can exist on a relatively dry and woody diet and can utilize habitats unavailable to cows. Some "hindgut" cellulose digesters subsequently ingest their fecal remains—"coprophagy"—to utilize many products that cannot be assimilated in the hind gut (e.g., rats, rabbits, grouse, etc.)

g. Increase the surface area with folds, rugae (smaller folds), and villi.

h. The "spiral" valve of the intestine in a shark, the typhlosole of the earthworm or lamprey gut, and the villi of the human intestine are examples of this strategy.

PART 2   The Role of Peristalsis

p. 40-7
a. The tongue is moved upward toward the roof of the mouth.

b. Yes, the movement would be the same and would movie materials toward the back of the mouth, forming a bolus—a soft, round mass of masticated food mixed with saliva.

c. The larynx moves upward and then down when he or she swallows.

d. When the larynx moves upward, the epiglottis moves to cover the passageway to the lungs.

e. If these movements did not occur, you could have food forced into the lungs.

5. 4 seconds.

p. 40-7
6. Rate = length/time = 2.75 inches per second .

f. The type of food, amount of water, pH, hypotonicity, as well as autonomic nervous system stimulation (stress, etc.) can affect rates of peristalsis.

EXERCISE B   The Chemistry of Digestion

PART 1   Carbohydrates

p. 40-9
a. Brownish-black. Undigested starch is present, but some starch breakdown has occurred.

b. Blue-black. I_2KI reacts with undigested starch.

c. Tube A_1 turns reddish-brown then gold; Tube A_2 remains blue-black. In Tube A_1, starch is gradually broken down from long chains of glucose (amylose and parts of amylopectin) to shorter glucose chains (dextrins) and eventually to the disaccharide maltose. When maltose is present, the color is gold. In Tube A_2, no amylase is present and the starch is not broken down so the color remains blue black.

Table 40B- 1

Tube	Enzyme	Substrate	I_2KI State	I_2KI Finish
A_1	5 ml amylase	5 ml starch	black	red-gold
A_2	5 ml H_2O	5 ml starch	blue-black	blue-black
B_1	5 ml maltose	5 ml A_1	red-gold	red-gold
B_2	5 ml H_2O	5 ml A_2	blue-black	blue-black

d. No glucose is present.

e. Amylase does not break down maltose. Maltose is broken down in the intestine by maltase.

f. Carbohydrates are long-chain polymers that are broken down sequentially, starting in the mouth and continuing in the small intestine.

g. The pH of the stomach is too low and enzymes designed for breakdown of carbohydrates will not work at this low pH.

Extending Your Investigation   Where Is It Digested? Enzymes and pH

p. 40-10   HYPOTHESIS: Amylase cannot digest starch at pH 2 as found in the stomach.

NULL HYPOTHESIS: Pepsin cannot digest proteins at pH 2 as found in the stomach.

Prediction—at pH 7, both enzymes will work, but at pH 2 neither enzyme will work.

PROCEDURE: Amylase (2g/100 ml) is prepared in solutions at pH 2, 4, 7, and 10 by adjusting the amylase solution with IN HCl or IN NaOH. Pepsin (50 g/1000 ml) is prepared in the same manner at the same pH values. Starch is used as a substrate (10% potato starch) for amylase and albumin (1%) is used as a substrate for pepsin. Add enzyme and substrate 1:1. Use I_2KI to test for the presence of starch and biuret reagent to test for the presence of proteins.

RESULTS:

	pH 3	pH 4	pH 7	pH 10
amylase	black	brown-black	red-gold	brown-black
pepsin	colorless to light violet	light blue-violet	violet	violet

Conclusion—amylase cannot work at a low pH; hypothesis supported. Pepsin can work at a low pH; hypothesis rejected.

PART 2   Proteins

p. 40-11   Table 40B-2 Action of Pepsin on Protein

Tubes	Contents			Results		
	Enzyme	Substrate	Additive	Film	Biuret	Ninhydrin
P_1	5 ml pepsin	5 ml albumin solution	2 drops 2 N HCl	clear dot	violet	purple
P_2	5 ml pepsin	5 ml albumin solution	2 drops NaOH*	opaque	violet	clear
P_3	5 ml H_2O	5 ml albumin solution	——	opaque	violet	clear
P_4	5 ml H_2O	5 ml albumin solution	2 drops 2 N HCl	opaque	violet	clear

*can be added to counteract low pH after addition of pepsin

a. P_1, yes. P_2, no. Albumin was partially digested by pepsin to form small polypeptides and dipeptides that still gave a violet color with biuret reagent. Since pepsin is itself a protein, all biuret tests yield a violet color.

b. The pH optimum for pepsin is very low.

c. Yes. Proteins were present in tubes P_2, P_3, and P_4, because pepsin was not breaking them down. Smaller peptides that are partial breakdown products were found in P_1 and also gave a positive biuret reaction.

d. Yes. Amino acids were only detected in tube P_1 in which the pH was low enough to allow pepsin to break proteins down into dipeptides. Heating for the ninhydrin test broke enough bonds to yield a purple color test for amino acids although pepsin does not break proteins all the way down into component amino acids.

e. Proteins are partially digested in the stomach to yield dipeptides (and poly-peptides, 3-10 residues). Digestion of proteins is then completed in the small intestine.

PART 3 Fats

p. 40-11 a. The water drops serve as a control.

p. 40-12 b. Bile salts.

c. Small fatty acids enter the blood vessels of the intestine directly, but larger fatty acids, glycerol, and cholesterol are processed differently. Within the cells of the intestine, fatty acids and glycerol are resynthesized to form fats which are then packaged into protein-coated droplets called chylomicrons. Cholesterol is packaged into low-density lipoproteion (LDL) complexes.

d. Both chylomicrons and cholesterol are secreted into lymph vessels and ultimately enter the circulatory system through thoracic ducts that enter veins in the chest. Chylomicrons break down to release fats to be stored in fat cells or fatty acids to be stored in muscle tissue. LDLs are picked up by the liver where cholesterol is stored, secreted in the bile, or repackaged for delivery to other body cells.

Table 40B-3

Substance	Enzyme/Breakdown Products/Location of Digestion or Absorption
Carbohydrates (polysaccharides)	amylase breaks down starch to dextrins in mouth and pancreatic amylase breaks dextrins into disaccharides in the small intestine
Disaccharides	disaccharases break disaccharides into monosaccharides in the small intestine; uptake of simple sugars in small intestine
Simple sugars	final carbohydrate breakdown product absorbed in small intestine
Proteins	breakdown in stomach (pepsin, HCl) and intestine (trypsin, carboxypeptidase from pancreas) to form small polypeptides and other protein fragments
Dipeptides	dipeptidases break these shorter products into amino acids in the small intestine; uptake of amino acids in the small intestine
Amino acids	final protein digestion product absorbed in small intestine
Fats	fats emulsified (bile salts) and digested (pancreatic amylase) in the small intestine to form fatty acids and glycerol; some partially hydrolyzed fats or colloidal "droplets" may pass directly through cell membranes to the lacteals of the lymph system within the intestinal villi
Fatty acids	uptake in the small intestine
Glycerol	uptake in the small intestine

p. 40-13   a. $C_6H_{12}O_6 \rightarrow 6\,CO_2 + 6\,H_2O +$ energy. The hydrogen of metabolic water comes from substrate molecules (simple sugars, fatty acids) and the oxygen from the atmospheric oxygen transported to tissues by the circulatory system.

PART 1   Anatomy of the Mammalian Kidney

p. 40-13   b.   They cushion the kidney and protect the fragile tubules and blood vessels of the kidney from mechanical damage.

c.   No.

d.   Active transport using ATP. Substances actively reabsorbed from the ultrafiltrate include glucose, amino acids, lactate, vitamins and most ions (including $Na^{+)}$ with most transport linked to that of Na^+. Moreover, the electrochemical gradient that drives most passive transport (Cl^-, water, etc.) is established by the active transport of sodium ions.

p. 40-14   **Figure 40C-1:** a, ureter; b, renal pelvis; c, renal papilla; d, nephron; e, medulla; f, cortex (e and f are general regions of the kidney, not specific structures); g, renal artery; h, renal vein.

p. 40-15   e.   ADH acts on cells lining the distal convoluted tubule and collecting duct.

f.   ADH controls the retention of water and maintains osmotic homeostasis. In the absence of ADH, water is *not* resorbed in these parts of the nephron and urine volume (with an attendant loss of electrolytes) increases to dangerous levels. This produces a condition called diabetes insipidis.

g.   Much of the water in urine is removed by the proximal convoluted tubule and descending loop of Henle. Active transport of sodium in the ascending loop of Henle retains much of the sodium found in the filtrate and recycles it within the medulla (countercurrnet multiplication) to produce and maintain an osmotic gradient between the medulla and cortex that is responsible for producing a hypertonic urine.

h.   Long, enabling the animal to resorb more water and produce more concentrated urine. Several species of rodents living in desert environments can subsist without drinking using metabolic water and moisture found in their food to survive and are, unlike us, able to drink seawater and survive. (Marine reptiles and birds have accessory salt-secreting glands, the supraorbital glands, that drain into the nasal passageways. Suggest to students that they watch a resting gull on their next visit to the ocean—they may see drops of concentrated, salt-containing water produced by these glands, forming at the tip of the bill of the bird has eaten recently.

Table 40C-1

Kidney Region	Structure	Function
Cortex	glomerulus	produce ultrafiltrate; the afferent arteriole is larger than the efferent arteriole so resistance and the hydrostatic pressure provided by the heart force many components of the blood into the surrounding capsule (blood cells and larger plasma proteins remain in the blood)
	renal capsule	collect ultrafiltrate; forms a functional unit with the glomerulus
	proximal tubule	active removal of sodium, glucose and other materials; water removed by diffusion
	distal tubule	active removal of sodium under control of aldosterone, an adrenal hormone; permeability of tubule lining to water controlled by a pituitary hormone, ADH; filtrate volume about 5% of original
Medulla	descending loop of Henle	passes through an osmotic gradient; volume of filtrate reduced to 15% of original during passage
	ascending loop of Henle	active transport of sodium; further reduction in filtrate volume
	collecting duct	passes through an osmotic gradient; volume of filtrate reduced to about 1% of original during passage; permeability of tubule lining to water also controlled by ADH

p. 40-16   Figure 40C-3

Note that water is not *actively* pumped from the kidney tubules—it follows osmotic gradients caused by the active transport of Na$^+$, particularly in the ascending loop of Henle. Note the role of ADH in altering the permeability of the distal convoluted tubule and collecting duct to water. In the absence of ADH, very large volumes of urine are produced (a condition known as diabetes insipidus).

p. 40-17   i. Salt-secreting cells are located on the gill lamellae of certain marine teleost fishes to remove excess sodium. Sharks possess a special segment in their kidney that retains urea which is used to keep their tissue fluids isotonic to the environment (with the rectal gland, a salt-secreting organ) regulating their ionic environment. Marine fish may also avoid drinking salt water, obtaining water from their food (whose body fluids may have lower tonicity than sea water) and their own metabolism. To conserve water, several lines of marine teleosts have also reduced the volume of filtrate produced in the kidney by reducing the size of the glomerulus. In some marine fishes, the glomerulus is missing altogether and the kidney is said to be "aglomerular" (the proximal tubule secretes some waste materials but osmoregulation is handled totally by the salt-secreting cells on the gill lamellae).

PART 2   Microscopic Anatomy of the Kidney

p. 40-17   a. Yes. Yes. However, in a normal microscope slide, these differences will be small and difficult for students to determine. The walls of collecting ducts will be thicker with more cells than those of the convoluted tubules. They should be about the same diameter as the ascending loops of Henle and will be difficult to distinguish from them.

PART 3   The Urogenital System

EXERCISE D   Gamete Formation

p. 40-18   **Figure 40D-1** (clockwise from upper right): primary spermatocyte (2*n*); secondary spermatocyte (*n*); spermatid (*n*); sperm (*n*); differentiation; meiosis II; meiosis I.

a. Spermatozoa production is continuous after puberty due to the continuos production of testosterone in humans and some our domestic animals that have been protected from the full force of natural selection. However, in many animals, reproductive activities are limited to one portion of an *annual cycle*. They are timed so that young are produced when adequate food is available for them to grow and survive (reproductive activities may be initiated by day-length changes—photoperiod—or, in populations living near the equator or in environments where the seasons are less predictable than in the higher latitudes, by other factors such as rainfall, the presence of green vegetation, presence or reproduction of prey populations, etc.)

b. The sperm travels through the seminiferous tubules, to the efferent ductules (rete testis), to the vas deferens (sperm duct). Sperm are stored in the epididymus (which develops from the coiled vas deferens) prior to ejaculation. Then the sperm travel through the vas deferens which empties into the urethra that extends through the shaft of the penis to the outside.

c. Oval.

d. Cell membrane, nucleus, mitochondria, and a small amount of cytoplasm.

p. 40-19   e. Sperm contain very little cytoplasm and therefore lack structures specialized for intracellular transport, such as endoplasmic reticulum, lysosomes, ribosomes.

f. Tubulin protein (mictotubules).

g. This process reduces the mass of the sperm and therefore improves locomotion.

h. The Golgi apparatus manufactures and packages protein macromolecules, including enzymes. Thus the acrosome is a modified Gogi vesicle containing lytic enzymes that assist the sperm in penetrating the protective structures that surround the egg.

i. The acrosome is crescent-shaped.

p. 40-20   **Figure 40D-4** (clockwise from upper right): primary oöcyte (2*n*); secondary oöcyte (*n*); first polar body (*n*); second polar bodies (*n*); ovum; biochemical differentiation; oötid (*n*); meiosis II; secondary oöcyte (*n*); meiosis I.

a. The corona radiata (follicle cells that surround the egg), the primary egg membrane or zona pellucida, and the egg cell plasmalemma. Primary membranes are produced within the ovary (zona pellucida, vitelline membrane, etc.) while secondary membranes (and shells), if present, are added in the oviduct. A chicken egg has a vitelline membrane (primary), layers of albumin, two shell membranes, and a shell (all secondary).

b. The egg could be fertilized outside of the oviducts and implant in the peritoneum.

c. Implantation of a zygote in the fallopian tubes.

p. 40-21   **Figure 40D-5** (a) antrumi (b) corona radiate (follicle cells); (c) ovum; (d) follicle cells; (e) zone pelucida.

Laboratory Review Questions and Problems

1. Materials are removed from the intestine through its surface—the more surface, the faster a given volume of digested products can be taken up. In lampreys, surface area is increased by a folded ridge projecting into the lumen (a "typhlosole"). In sharks and several other fishes, a "spiral" or valvular intestine increases the surface area. Most vertebrates, however, lengthen and fold the small intestine to increase the surface area. Within the intestine, ridges or rugae may appear and, in mammals, finger-like villi further increase the surface projecting into the lumen. Microvilli on the exposed surface of the mucosal cells also increase the area for uptake. Animals that eat less readily digested foods (plants, etc.) often have longer and more massive small intestines than those that feed on other animals or more readily digested materials.

2. The liver produces bile, which is added to the intestinal contents. Bile contains bile pigments, waste products of hemoglobin metabolism (these give feces their characteristic dark color), and bile salts which aid in breaking up and emulsifying fats so that they can be attacked by digestive enzymes or assimilated directly by the cells lining the intestine. The liver also receives blood from the intestines through the hepatic portal system. This blood contains most of the end products of digestion. Most of these are treated in the liver. Simple sugars are stored as glycogen in the liver and other body tissues. Amino acids are broken down into carbohydrates or fatty acids with the production of nitrogenous wastes which are converted to less soluble urea (or insoluble uric acid in some animals) in the liver. Fatty acids are packaged for transport and shipped to adipose tissues. Many waste products are broken down or detoxified in the liver (the liver is damaged by excessive and chronic consumption of alcohol exceeding its normal capacity for treatment) . Blood proteins and many other products are synthesized by the liver. The organ is necessary for life in all vertebrates.

3. For added detail on digestion within the small intestine, see Figure 40B-1.

Region	Treatment	Secretions	Products
mouth	chewing; mix with saliva (mucus provides lubrication)	amylase	dissacharides
esophagus	transport	none	none
stomach	mix with gastric secretions	HCl, acid; pepsin	break down proteins
small intestine	mix with mucosal products and secretions from the pancreas and liver	bicarbonate, many enzymes	break down carbohydrates, proteins, and fats
large intestine	resorb water	none	none

4. Bowman's capsule and the convoluted tubules are located in the cortex toward the outer surface of the kidney. The loops of Henle plunge into the medulla where a steep osmotic gradient helps remove water and active transport removes and retains sodium to maintain the cortical-medullary gradient. The collecting ducts carry urine into the pelvis where it leaves by the ureter for transport to and storage in the urinary bladder.

5. In mammals, much of the volume of blood passing through the glomerulus is filtered, removing water and low molecular weight compounds but retaining cells and osmotically active large proteins. Blood leaving the glomerulus enters vessels that course into the medulla and bathe the loop of Henle where they are involved in maintaining osmotic gradients within the kidney and removing recovered materials such as water, salt, and glucose. H^+ ions, amino acids, bicarbonate, and carbonic acids are also filtered in the proximal convoluted tubule, affecting pH of the blood in capillaries surrounding the nephrons and in the interstitial fluids.

6. The urogenital system is composed of elements that produce urine (excretory and osmoregulatory functions are involved) and elements that produce gametes (eggs and sperm). Despite the different functions of these systems, they develop from tissues of the mesomere in the embryo and share certain structures as a result of this proximity. For example, in mammals, a new kidney duct (the ureter) appears during development and, in males, the embryonic duct becomes modified to carry sperm as the vas deferens in the adult.

7.

	Structures Formed During:	
Division Process:	**Spermatogenesis**	**Oögenesis**
Meiosis I	secondary spermatocyte	secondary oöcyte
Meiosis II	spermatid	oötid
Differentiation	sperm	ovum

8. Unequal division forms a large secondary oöcyte ($1n$), which contains almost all of the cytoplasm, and smaller polar bodies which degenerate.

More:

Laboratory 41

Control—The Nervous System

PART I   **Neurons**

EXERCISE A   Examining Nerve Cells (Neurons)

p. 40-2   **a.** The cell body (soma) contains the nucleus and most cellular organelles. The axon conducts nerve impulses away from the cell body. Dendrites conduct nerve impulses toward the cell body (some authors may restrict the term to the terminal receptive or input region in sensory neurons and regard all conducting fibers as axons). Use these terms in agreement with your text. Note that the *direction* in which action potentials are conducted is not a property of the neuron but of the synapse—the cell-cell junction at which neurotransmitters (pre-synaptic) and their receptors (post-synaptic) gate or polarize the nervous system functionally.

b. Sensory (afferent) neurons receive sensory information (usually transduced by an epithelial element) and relay it to the central nervous system. Interneurons in the gray matter of the brain transmit signals within localized regions of the central nervous system. Sensory and interneurons are spontaneously active but motor neurons respond only to adequate stimulation. Motor (efferent) neurons transmit signals from the central nervous system to effecters (muscles, glands, etc.).

c. Nerves are organs composed of bundles of fibers of many neurons (conducting elements of several sizes). Because it is not possible to identify functional dendrites or axons in a nerve (unless all are regarded as axons), these cellular processes are often referred to simply as fibers. Nerves are surrounded by a connective tissue sheath that is really quite strong.

d. Stimulated by the arrival of an action potential, neurotransmitters are released presynaptically, diffuse across the synaptic cleft, and combine with receptor molecules on the postsynaptic membrane to produce a propagation of the nerve impulse. The synapse is polarized (it transmits information in one direction only) because of this difference in structure of the two cells. (Note that single pre-synaptic events are rarely if ever transmitted postsynaptically. The "decision" to pass information along is determined by the input activity—both inhibitory as well as stimulatory—of thousands of axons impinging on the particular neuron with different temporal and spatial patterns of activity. In the special case of neuromuscular junctions, the motor fiber forms multiple synaptic junctions so the effects of a single action potential are summed spatially and propagate a single action potential to the muscle cell.)

PART II   Sensory Receptors

p. 41-3   **a.** Receptors for different taste sensations are located in different areas of the tongue.
b. No taste.

p. 41-4   **Figure 41B-2** *Localization of taste sensations on the tongue*

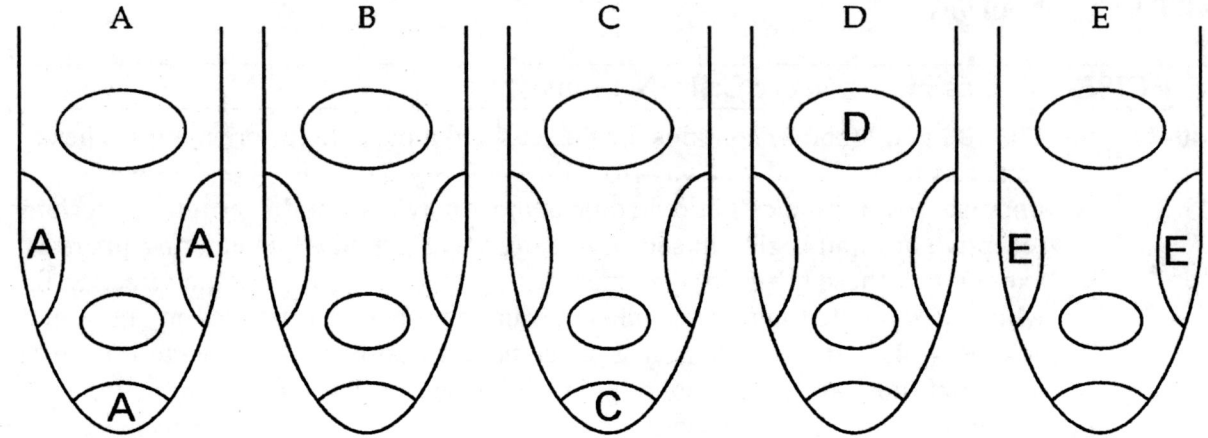

	A	**B**	**C**	**D**	**E**
taste	salty	none	sweet	bitter	sour
solution	sodium chloride	water	sucrose	quinine	acetic acid

c. Bitter.
d. Different receptors for various types of taste sensations are located in different areas of the tongue. The stimulating material must contact the correct sensors for that material.

EXERCISE C   Chemoreceptors: Individual Differences in Taste

p. 41-4   HYPOTHESIS: If taste perceptions differ from person to person, then there will be a difference in the way taste papers and certain foods taste to one person as compared to another.

·   NULL HYPOTHESIS: There will be no differences in the way taste papers and certain foods taste to different individuals.

a. Prediction—individuals will differ in their taste responses to taste test papers.

p. 41-5   Independent variable—different substances tested for their taste
Dependent variable—taste sensation

RESULTS:

control (white)	tasteless
PTC (blue)	bitter or tasteless (the difference is genetic)
thiourea (yellow)	bitter
sodium benzoate (pink)	tasteless to some individuals; sweet, sour, bitter, or salty to some individuals

    b. Variable.

    c. No.

    d. Sodium benzoate tasted either sweet, sour, salty, or bitter to various members of the class.

    e. Sodium benzoate would cause the food to taste differently (sweet, sour, bitter, or salty) to different individuals.

    f. Relate your answer to the previous paragraphs and tastes recorded under 2, p. 41-5.

    g. Observations support the hypothesis but not the null hypothesis.

    h. Conclusion—chemoreception and food preferences differ in different individuals. Friends who tasted PTC as bitter liked foods that were not liked by nontasters. Friends who tasted sodium benzoate as salty liked sour foods better than those who tasted sodium benzoate as bitter. The conclusion is that there are either differences in different individual's chemoreceptors or in the way information from chemoreceptors is processed by the brain that are responsible for liking or not liking certain foods. Differences are probably genetic, but this would be an inference based on observations.

EXERCISE D   Chemoreception: Smell Discrimination and Its Influence on Taste

p. 41-6   HYPOTHESIS: If you cannot smell the food you taste, it's taste will be different from the taste of food that has been "smelled."

    NULL HYPOTHESIS: Food will taste no different whether it stimulates only taste receptors on the tongue or these receptors plus olfactory receptors of the nose.

    a. Prediction—you cannot properly identify foods if sampled without being allowed to smell them.

    Independent variable—the material presented for tasting.

    Dependent variable—the taste sensation.

p. 41-7   5. Flavor with nose held—cannot tell, fruity.

    Flavor without holding nose—cherry.

    Actual flavor cherry.

    b. Smell affects the taste of Life Savers.

    c. Results support the hypothesis but not the null hypothesis.

    d. Olfactory (smell) and gustatory (taste) sensations are integrated by the brain to produce a complex perception of the "taste" of a substance. Foods often seem to be "tasteless" when one has a cold because the nasal passages are blocked and olfactory sensations are reduced or missing.

EXERCISE E   Photoreception: Vision—Structure of the eye

p. 41-7,8   **Figure 41C-1** *Structure of the eye*

    (clockwise from upper right): b, sclera; d, choroid; k, retina; a, optic nerve; j, vitreous chamber; f, pupil; i, aqueous chamber; c, cornea; e, iris; g, lens; h, ciliary body. Not shown—extrinsic eye mussels. The part of the aqueous chamber anterior to the iris is the anterior chamber, that posterior to the iris is the posterior chamber. Both are filled with aqueous humor. The vitreous chamber is filled with vitreous humor.

---

p. 41-9 4.

Outer Layer	Middle Layer	Inner Layer
sclera, cornea	choroid (includes iris, suspensory ligament, and ciliary muscles)	retina—sensory elements (rods and cones); pigment layer; part of iris

Note that the conjunctiva is a layer of the skin and not part of the eye itself.

a. Light passes through the conjunctiva (skin), cornea, aqueous chamber, pupil, lens, and vitreous chamber to the retina.

IRIS

EXERCISE F   Photoreception: How We See

p. 41-9 As students study Figure 41F-1, you might point out that the epithelial sensory elements of the eye are the rods and cones. The first order neuron, the bipolar cell, is really the optic "nerve" and the second order neuron that carries information to the brain through the so-called optic nerve, the ganglion cell, is really part of the brain—a tract, not a nerve. In fact, the eye develops as an evagination of the diencephalon that gives rise to the retina and pigment layer.

a. The fovea

b. Cones are concentrated in this area and form a 1:1 connection with bipolar cells—toward the periphery, resolution is sacrificed as several sensory elements connect to a single bipolar cell but sensitivity is increased by the summed input.

c. Look out of the side of your eye past the object you are trying to see.

d. Rods are much more sensitive to light and respond when it is dark. In fact, rods are inactive in bright light (their pigments are all bleached). Rods are also located toward the periphery of the retina. Because many connect with one bipolar cell, the image will not, however, be sharply focused.

PART 1 Peripheral Vision and Color Vision

p. 41-11 **a.** The shape should be determined first, since the rods on the periphery of the retina would be the first to perceive an object coming into view from the side and the rods are insensitive to color. Color would be determined only when the object was sufficiently in front of the eye to be perceived by cones.

PART 2 The Blind Spot

p. 41-11 **a.** Each eye compensates for the other eye's blind spot (the blind spot of the right eye blocks off a different part of the image than does the blind spot of the left eye).

b. The brain "fills-in" an image using information from the surrounding region.

EXERCISE G   Mechanoreception: The Role of Sensory Receptors in Touch

p. 41-12 2. Sample data: lips, 2-5 mm; fingertips, 3-5 mm; palm, 12 mm; back of hand, 12 mm; inner wrist, 4-9 mm; outer wrist, 6-15 mm; inner forearm, 18-42 mm; outer forearm, 24-100 mm

a. The fingertips are your first line of "touching defense" and are extremely sensitive—your skin has have very large number of Meissner's corpuscles in the fingertips.

b. Yes. The nose and lips are areas of the face that would first encounter stimuli. For

---

this reason, they should have a large number of Meissner's corpuscles.

    c. The lips and fingertips were most sensitive to light touch. The most sensitive areas (at the anterior end of the body and on the distal appendages) are those which would ordinarily first encounter stimuli and need to be capable of greater discrimination.

EXERCISE H   Thermoreception: Discriminating Temperature

p. 41-13   **a.** The finger in hot water feels hot because only the warm receptors are stimulated. The finger in cold water feels cold because only the cold receptors are stimulated.

    b. Both fingers are comfortable and no longer experience the temperature extremes as they have become habituated to the water temperatures in the beakers.

    c. At room temperature, both warm and cold thermoreceptors are stimulated. The finger displaced from the cold water beaker feels warmer when placed in the warm water. The finger displaced from the hot water beaker feels cooler.

    d. The stimulus for both fingers is the warm water that stimulates the cold receptors in the finger that had been in warm water and the warm receptors in the finger that had been in cold water.

    e. Heat gain.

    f. Heat loss.

EXERCISE I   Proprioception: The Role of Proprioceptors in Determining Position

p. 41-13   **a.** The fingers did not touch.

    b. The fingers may touch behind the back, though it is still difficult. Vision plays an important role in determining position and regulating muscle activities for any action occurring within the visual range. The body is less accustomed to using vision for actions behind the back.

    c. Sensors in the fingers and proprioceptors in the joints of the fingers allow the brain to integrate information about position of the string and fingers with knowledge (memory) of the visual steps associated with tying knots.

    d. Information from both joints (stretch receptors) and muscles (muscle spindles) is probably involved. Pressure and touch receptors in the dermis assist in this process.

EXERCISE J   Mechanoreceptors of the Ear

p. 41-15   **a.** The eyes move opposite to the direction (spin) of the body movement.

    b. Movement of fluid in the semicircular-canals stimulates the hair cells of the ampulla. Stimulation is minimal when velocity is constant and maximal after sudden acceleration or deceleration because the endolymph is not tightly coupled to the wall of the canals. After a period of rotation at constant velocity, the body and the fluid come to rest with respect to each other, but with a deceleration (or further acceleration) a relative movement is sensed. This gives us information about our changing position in three-dimensional space. This provides a dynamic position sense as opposed to the static position sense provided by other sensors in the inner ear not affected by movements of the endolymph.

    c. Turning the head in the direction opposite to that in which it is spinning minimizes pressure in the ampulla of a semicircular canal and thus minimizes dizziness.

    d. Because your movements are not steady—acceleration and deceleration occur continuously and stimulate the sensors in your inner ear. If movements were steady (as in a centrifuge), you would not become dizzy

PART III   Examining the Structure of the Brain

EXERCISE K   The Structure of the Mammalian Brain

p. 41-15 a. The following answers pertain the mammalian brain (in sub-mammalian vertebrates, major sensory perceptions, in particular, may be mediated in separate regions of the brain):

controlling body temperature	hypothalamus (diencephalon)
coordinating equilibrium	cerebellum (rhombencephalon; metencephalon)
controlling breathing	medulla oblongata (rhombencephalon)
intelligence	cerebral hemispheres (telencephalon)
sensory perception	cerebral hemispheres (telencephalon
heart rate	medulla oblongata (rhombencephalon)
the relay of impulses between lower brain centers and the cerebrum (mammals)	thalamus (diencephalon)

p. 41-16 appears to the left of "controlling breathing".

p. 41-16 Figure 41K-1 *The sheep's brain*

clockwise from the top: b. corpus callosum; c. thalamus; d. epithalamus; f. tectum; a. cerebrum (cerebral hemispheres); g. cerebellum; i. medulla oblongata; j. choroid plexus (IV ventricle); h. pons; e. hypothalamus.

Laboratory Review Questions and Problems

1. Sensors monitor an animal's internal physiology and external world (internal and external environments). Effectors (bones, muscles, glands) produce appropriate actions as responses based upon the effects of stimuli interacting with genetic reaction norms, structural capabilities, and experience (learning).

2. Neurons are the functional cells that make up the nervous system. Synapses are their junctions with other nerve cells (cell-cell junctions). Neurotransmitters are released across synaptic clefts as the result of the arrival of an action potential and control the direction of transmission and the particular effect of a nerve impulse. Therefore, the characteristics of the neurotransmitters at a synapse control the end results of nerve impulses.

3. Both are perceived by chemoreceptors and depend upon the effects of the stimulating molecule(s) upon epithelial receptor cells. Olfactory receptors of the nose are sensitive to lower concentrations of molecules and to a greater range of perceived odors than are taste buds of the tongue (they may be able to produce a "sensory" event in the presence of a single molecule). Olfactory sensors are nerve cells that give rise directly to the olfactory "nerve" fibers leading to the brain (and are replaced throughout much of the life of many vertebrates, unlike most nerve cells). Gustatory sensors are epithelial cells associated with sensory nerve

cells. They too are replaced when damaged.

4. Mechanoreceptors in the inner ear (semicircular canals and statoreceptors or maculae acousticae) detect changes in the motion and position of the head. The cerebellum integrates this information with sensory input from proprioceptors.

5. In mammals:

> vision—hypothalamus via the lateral geniculate bodies to the cerebral cortex and midbrain tectum
>
> smell—olfactory bulbs to olfactory lobes of the telencephalon
>
> hearing—to the midbrain tectum and through the thalamus (medial geniculate bodies) to the cerebral cortex
>
> thought and voluntary behaviors—cerebral hemispheres
>
> visceral functions and control of the anterior pituitary gland—brain stem, including the hypothalamus
>
> respiration, heart rate, vasoconstricdon, and vasodilation—medulla oblongata (part of the rhombencephalon)
>
> equilibrium, posture, and fine control of voluntary movements—cerebellum
>
> speech—cerebral cortex

Notes:

Laboratory 42

Behavior

EXERCISE A	Reactions of Isopods to Light and Humidity

p. 42-2   1.   HYPOTHESIS: Pill bugs will congregate in darker environments, particularly under the lid with the moist paper towel (higher humidity).

NULL HYPOTHESIS: Pill bugs will be randomly distributed within their environment (without regard to the lids or differences in humidity between the lids).

Prediction—more isopods will be under the lid over the moist paper towel with higher humidity than under the other lid over a dry towel or wandering outside either lid.

Independent variables—light and humidity (moist vs. dry environment).

Dependent variable—movements of isopods in response to light and humidity.

p. 42-3   Number wandering: <u>4</u>. Number in moist habitat: <u>6</u>. Number in dry habitat: <u>0</u>.

Results support the hypothesis but not the null hypothesis.

Conclusion—isopods demonstrate behavioral adaptations which lead them to moist habitats.

2.   HYPOTHESIS: In darkness, pill bugs will gather in the moister environment.

NULL HYPOTHESIS: Pill bugs will be randomly distributed between the drier and moister environments in darkness.

Prediction—more isopods will be under the lid over the moist paper towel with higher humidity.

Independent variable—humidity (moist vs. dry environment).

Dependent variable—movements of isopods in response to humidity.

Number wandering: <u>3</u>. Number in moist habitat: <u>7</u>. Number in dry habitat: <u>0</u>.

Yes, the hypothesis (but not the null hypothesis) is supported.

a.   Pill bugs are negatively phototactic (move away from light). Movement is random until the pill bugs find a moist environment, at which time their activity level slows and the pill bugs remain within the moist environment.

b.   No, the results in light and in dark are very similar. However, both lids produce a dark refuge in both of these experiments.

3.   HYPOTHESIS: Pill bugs will congregate under the opaque paper box top rather than under the clear plastic top.

NULL HYPOTHESIS: Pill bugs will be randomly distributed between the clear and opaque refuges.

Independent variable—light.

Dependent variable—movements of isopods in response to light

Number wandering: <u>6</u>. Number in light habitat: <u>0</u>. Number in dark habitat: <u>4</u>.

Results support the hypothesis but not the null hypothesis.

p. 42-4   4.   Number wandering: <u>2</u>. Number in light habitat: <u>3</u>. Number in dark habitat: <u>5</u>.

c. Isopods demonstrate behavioral adaptations which lead them to dark habitats.

d. Yes, isopods move to the dark.

e. The combination of humidity and darkness enhances the eventual settling of the pill bugs in the dark habitat. The pill bugs move away from light, thus the dark habitat is preferred. The moisture in the (moist) dark habitat results in a slowing of the pill bugs activity level, so they remain within the dark habitat.

5. HYPOTHESIS: Isopods will prefer the dark-dry habitat to the moist-light habitat.
   NULL HYPOTHESIS: Pill bugs will be randomly distributed among the refuges and the general environment—they will exhibit no preferences for any of the physical conditions.
   Prediction—Pill bugs will congregate in the dark-dry habitat.
   Independent variables—light-moist vs. dark-dry conditions
   Dependent variable—movements of the pill bugs
   Number wandering: 3. Number in moist-light habitat: 3. Number in dry-dark habitat: 4.
   The isopods show no clear preferences, supporting the null hypothesis but not the hypothesis.

f. Taxis with respect to light. Kinesis with respect to moisture. The pill bugs wander randomly due to an increased activity level as long as they are in a dry habitat, and orient themselves in a direction away from light.

g. Steps 3, 4, (and 5).

h. Steps 1, 2, (and 5).

g. The most wandering was observed in the experiment that lacked a moist habitat. The increased activity level of the pill bugs caused this wandering.

h. Photoreceptors (ocelli) are sensitive to the level of illumination. Humidity information may be obtained via plate-like aesthetascs believed to be chemoreceptors. Avoidance of dry conditions is important because isopods do not have a waxy cuticle to retard desiccation and continue to use gills for gas exchange (some have pseudotracheae and can tolerate dry conditions). Their tendency to roll into a ball (hence the name "rolly-polly") is probably a defense against water loss through the ventral exoskeleton.

p. 42-6   Table 42B-1 Sample Table for Recording Observations of *Drosophila* Behavior

Time		Mating	Mating	Mating	Unreceptive Female	Two Males
Min.	Sec.	Pair A	Pair B	Pair C	Pair D	Pair E
01	10	wandering	wandering	wandering	wandering	wandering
	20	wandering	wandering	O	wandering	wandering
	30	wandering	wandering	O	wandering	O
	40	wandering	T	T	wandering	O
	50	wandering	O	V	O	T
	60	O	O	V	O	V
02	10	O	V	L	O	V
	20	O	V	decamping	O	O
	30	V	S	O	decamping	O
	40	V	L	O	decamping	O
	50	V	depressing	V	decamping	decamping
	60	O	O	V	O	decamping
03	10	O	O	V	O	decamping
	20	O	L	L	V	decamping
	30	0	decamping	AT	V	decamping
	40	V	O	decamping	decamping	wandering
	50	V	O	decamping	O	wandering
	60	L	V	decamping	V	wandering
04	10	L	V	O	L	wandering
	20	AT	V	V	decamping	
	30	ignoring	L	L	V	
	40	L	AT	L	V	
	50	L	L	AT	L, AT	
	60	decamping	AT	C	decamping	
05	10	AT	C	decamping		
	20	O	decamping			
	30	O	decamping			
	40	V	decamping			
	50	L, AT	decamping			
	60	C	decamping			

p. 42-7

a. The shortest time was 2 minutes, the longest, 4.5 minutes. The average was 3.2 minutes.

b. A submissive female spreads her wings and extends her genitalia.

c. The male takes the more active role in courtship.

d. Often more than one male would orient toward and circle females.

e. Place flies in a plastic tube, darkened at one end with tape. Shine a light on the tube and determine if photoreceptors are-responsive during courtship and mating.

PART 1   Aggressive Behavior and Social Dominance

p. 42-9   Table 45C-2 Aggressive Encounters

Pairing	Level 1	Level 2	Level 3	Level 4	Level 5
Y vs. G	6	YYY	YGYYY	YYG	
Y vs. B	5	YBYBY	YY	YBBY	Y
Y vs. R	3	YYYRR	YYYR	RR	
G vs. B	5	GBG	GBB	BBGGG	BG
G vs. R	3	GRRR		GRR	RG
R vs. B	7	RRRBRR	RBBRR		R

Yellow	wins 1, losses 2
Green	wins 2, losses 1
Blue	wins 3, losses 0
Red	wins 0, losses 3

p. 42-10   Table 42C-3 Use of Cricket Houses

Pairing	Cricket Using House
Y vs. G	YYYG
Y vs. B	YB
Y vs. R	YRRYR
G vs. B	GRRG
G vs. R	RGR
R vs. B	RRB

a. Yes.

p. 42-11   Table 42C-4 Cricket Dominance

Cricket	Total Point Score
Yellow	143
Green	138
Blue	146
Red	134

blue → yellow → green → red

b. They touch antennae or one cricket runs its antennae over the body parts of the other. If the second cricket does not retreat, an aggressive encounter usually ensues. The crickets lash their antennae around and posture by lifting the rear of their bodies. They sometimes kick with their back legs and shake the entire body. Stridulations become louder and, once one male starts, the other also stridulates, competing in loudness. They rush forward, butting heads and opening their mandibles. They grapple, wrestle, bite, and may throw each other sideways or flip over. After an encounter, the dominant male usually stridulates or chirps, but the subordinate male usually does not make any noise.

c. Yes. They become louder.

d. Wings.

e. Yes.

    f. Yes. Encounters appear to be more aggressive when they approach each other from the head end.

    g. Yes.

    h. 10 minutes.

    i. Dominant.

PART 2 **Aggressive Behavior, Dominance, and Courtship Among Male and Female Crickets**

p. 42-12 **Table 42C-8 Courtship Dominance**

Level 1	Level 2	Level 3	Level 4	Level 5
MMF	M	MF	M	M

    a. Male. Female crickets respond to chirping of males; the female moves to the stationary male. If a male and female meet head on, the female usually becomes immobile. This elicits courting behavior rather than aggression. A super-aggressive male may force a female to retreat.

    b. Yes. The sound of the stridulations changed. The females mounted the males from the posterior end.

    c. No. The female has auditory organs on the tibia of the front legs and are very accurate in determining from which direction a male's song is coming.

    d. The front margin of the forewing acts as a scraper and is rubbed over a "file" formed by a vein in the forewing. The wings cross over and one wing acts as the scraper while the other acts as the file.

    e. Yes. They were raised in a tent-like formation over the abdomen.

    f. The song changes from a loud calling song to a softer courtship song.

    g. Body posture, locomotion, touching of antennae and body parts.

    h. Yes.

EXERCISE D   Learning in the Mealworm

p. 42-14 **a.** Other stimuli include: food or moisture (positive reinforcements); excessive heat or a chemical deterrent (negative reinforcements).

    b. Less. The mealworm is less intelligent (has a less sophisticated mechanism for responding to stimuli and learning).

Laboratory Review Questions and Problems

1. A taxis is more efficient in leading animals to congregate in an optimal habitat, since it involves sensing of a gradient. A kinesis is most useful in causing animals to remain within an optimal habitat, once they have found it, and does not necessarily assist the animals in locating the optimal habitat.

2. Lack of water would present a motivation which may increase the activity level of the organism as it searches for a water source (until the organism becomes severely dehydrated, at which point activity level will slow). When presented with water, the animal will drink. The increased activity as the animal searches for water is appetitive behavior. The drinking of water is consummatory behavior.

3. **Habituation** (a response to a repeated stimulus): sleep cycles, meal times, ignoring traffic outside your window.

   Conditioning (a learning style in which one stimulus ("conditional") is associated with and elicits the same response as another ("unconditional" stimulus). An example is answering the telephone when it rings.

   Insight learning (solving a problem without prior practice): a baby trying to get out of a crib uses toys or pillows to lift her body up higher to get over the rails.

4. The advantage in relation to learning is that animals would have familiar areas in which to hunt for food or hide. It is a range in which they could have explored all of the potentials of the environment and would know where resources are located.

5. Although humans have been "socialized," this type of competition still occurs in sports, competition for grades in school, and even during physical disasters where supplies and shelter are in short supply.

6. Performance of the ritual and interactions with another participant (or participants) aid in bringing both male and female to physiological and behavioral readiness to mate. The exchange of species-specific signals also ensures that participants belong to the same species and do not waste time or gametes in unfruitful unions. Individuals are assured that they have ready partners of the same species. The species is continued by the combined reproductive activities of its members.

Laboratory 43

Communities and Ecosystems

p. 43-2   **Table 43A-1 Raw Data Table for __1__ Week-Old Culture**

	Number of organisms in 0.01 ml cluture medium		
	(A) *P. aurelia*	(B) *P. caudatum*	(C) *P. caudatum/P. aurelia*
Sample 1	44	54	7/20
Sample 2	60	58	8/30
Average	52	56	8/25

p. 43-3   **Table 43A-2. Density Estimates: Summary of Data for Entire Class**

	Grand average number of organisms in 0.01 ml of culture medium			
	Starting density	One week	Two weeks	Three weeks
Pure cultures:				
P. aurelia	1-5	52	56	54
P. caudatum	1-5	19	21	18
Mixed culture:				
P. caudatum	1-5	8	6	2
P. aurelia	1-5	25	35	37

a. *P. aurelia* had a greater increase in numbers in pure culture than did *P. caudatum*.

b. In the presence of a competitor, the growth rate of *P. aurelia* decreased. *P. aurelia* growth was much more rapid than was growth of *P. caudatum* because *P. aurelia* was able to utilize the food source more efficiently.

c. *P. caudatum* was driven to virtual extinction by the end of three weeks. *P. aurelia* *is* the better competitor.

8.

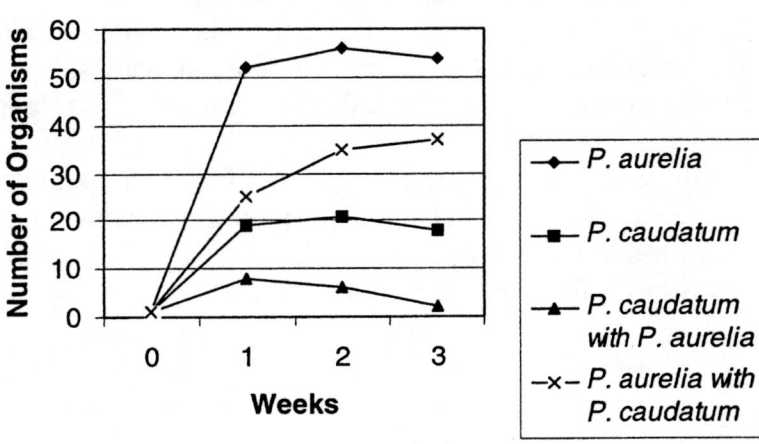

Species Competition

p. 43-5   Table 43B-2 Invertebrates identified by individual students

Phylum	Class	Number of organisms collected		
		Woodland	Riparian	Forest
Nematoda		0	1	1
Annelida	Oligochaeta	1	1	0
Mollusca	Gastropoda	3	5	2
Arthropods	Insecta	15	12	15
	Arachnida	9	6	7
	Crustacea	5	0	3
	Diplopoda	2	0	2
	Chilopoda	2	1	2
Other		0	0	0

Table 43B-3 Class data for all invertebrates collected

Phylum	Class	Number of organisms collected		
		Woodland (n_i)	Riparian (n_i)	Forest (n_i)
Nematoda		0	27	27
Annelida	Oligochaeta	23	26	7
Mollusca	Gastropoda	72	139	58
Arthropods	Insecta	384	306	386
	Arachnida	219	138	178
	Crustacea	118	0	85
	Diplopoda	37	0	40
	Chilopoda	41	14	38
Other		0	0	0
Total (N)		763	686	819

p. 43-6   5. Shannon Indices:

Woodland: 0.617 .   riparian: 0.649 ; forest: 0.677 .

a. Greatest diversity: Woodland Insecta ; riparian Insecta ; forest Insecta .
Least diversity: Woodland Nematoda and Oligochaeta ; riparian Crustacea
· and Diplopoda ; forest Nematoda and Oligochaeta. Note that soil nematodes are
probably underrepresented in these samples.

b. The forest (mature hardwood forest) had the greatest number of classes. The
riparian environment had the least number of classes.

c. The forest had the greatest number of organisms. The riparian habitat had the least
number of organisms.

d. The diversity and quantity of organisms in the leaf litter communities were
enhanced in moist (high humidity) environments having ample and long-standing
litter for decomposition.

e. The forest is the most mature ecosystem.

EXERCISE C   Using Climate Data as an Index to Vegetation

p. 43-7   Note: Weather data for world cities is in the *Preparator's Guide.*

p. 43-8   **a.**   Singapore (9.3 - 67.0 cm). However, Mondou, Chad exhibits annual extremes in rainfall which are almost as great (0 - 45 cm) and span a more critical range.

b.   In the Singapore tropical forest, lacking extreme variations in temperature, the major adaptations of animals to the excessive rains are behavioral (tree-dwelling existence). In the Savannas of Chad, the extreme annual rainy season/dry season cycle is reflected in the growth cycles and behavioral adaptations of the endemic plants and animals. The aboveground portions of many of the plants die, or are leafless during the dry seasons, sprouting again from the roots during the rainy seasons. The animals adapt to the extremes of water availability by metabolic adaptations that allow body functioning with relatively little water, and behavioral adaptations, including migrations to reach water sources.

c.   Kalgorlie, Australia (desert).

d.   Many desert plants condense their entire life cycles (vegetative growth, flowering, and seed production) into the brief periods of time when water is available. Other, large plants are adapted to the dry conditions by production of leaves only when water is available, reduction of total leaf area to conserve water loss (cacti are an extreme example: leaves are reduced to spines and the stems are photosynthetic), or have photosynthetic cycles (CAM, C_4) which minimize water loss. Desert animals are adapted to the hot dry conditions due to their capability to conserve water and, as in many small desert animals, nocturnal activity that allows them to avoid the extreme midday heat. Some desert animals may aestivate during the driest months.

e.   The coldest monthly temperature is found in the tundra, but the greatest change during the year is characteristic of the taiga. Latitudinal tundra is sufficiently far north that the sun's rays strike the earth more obliquely even in summer—however, it has very low precipitation throughout the year and temperature extremes are moderated by proximity to water. Much of the taiga is located within continental land masses where winter temperatures can be quite extreme. Occurring at lower latitude, however, summer temperatures are warmer than those found in tundra.

f.   At temperatures below freezing, plants become dormant or loose their leaves. Those unable to endure freezing survive as seeds or other propagules. Ectothermic animals become inactive or survive as eggs, pupae, or in some other resting state. Endothermic animals may remain active, foraging on plant foliage (ptarmigan, snowshoe hares, muskoxen) or other animals (foxes, polar bears), but many either migrate (caribou, many birds) or hibernate (lemmings) to escape the harshest part of the winter.

Laboratory Review Questions and Problems

1.   A more complex habitat may decrease the intensity of the competitive interactions between two species. A complex environment may allow the less competitive species to find food in areas sheltered from the more competitive organism. In cases where one of the organisms preys upon the other, a complex environment will provide shelter from predation. A more

complex environment will alter the growth curves for the two competing species by prolonging the existence of the less competitive species.

2. You would find more intense competitive interactions among the more closely related species, which are likely to share similar requirements for food, living conditions, and life style. Competition among sympatric species that are closely related may exaggerate differences (on islands of the Galapagos where closely related species of Darwin's finches overlap, changes in bill size enable the two competitors to utilize different food resources). This phenomenon is known as character displacement.

3. As the overall diversity of a community increases, physical structures (layers, patchiness, etc.) diversify and the number of available ecological niches increases. More mature communities contain a greater diversity of niches.

4. Temperature, incident light and associated environmental variables such as the length of the growing season and the availability of food supplies all decrease along this transect. Thus the overall diversity of the communities also decreases along the transect as the number of available niches is reduced.

    The pattern of rainfall is, however, more complex. In general, it also follows a gradient from the equator to the poles, but at about 30° N and S latitude, the warm equatorial air masses, which give up their moisture near the equator as they rise and cool, descend. They contain little moisture and this region is characterized by low rainfall and desert conditions around the globe (our Sonoran desert, the Australian interior, the Sahara desert, etc.). Farther north, convective uplift provides moisture distributed less evenly throughout the year by continental weather patterns. Monsoons also affect rainfall in many parts of the world— during the summer, continents are warmer than their surrounding oceans so warmed air rises and brings cooler air and moisture over land with heavy rains; during winter, continents are colder than the surrounding oceans and dry, cool air prevails. In Mediterranean climates, summers are hot and dry, winters are cool and wet. Fire is also an important ecological factor in these biomes (and others with seasonal rainfall).

5. The changes in community structure as you climb a tall mountain situated near the equator are similar to those you would encounter if you traveled towards the poles. The change in atmospheric temperature as you climb each 100 m corresponds to that you would experience by increasing latitude by 1° towards the poles. You would thus traverse from the hot, humid tropical rain forests at the base of the mountain, through a region of temperate forests, through a taiga zone dominated by conifers. Above a "tree line," an excessively cold climate and a short growing season, combine with intense winds to inhibit tree growth. Thus the vegetation on the upper slopes of very tall mountains may resemble that of the tundra (a treeless grassland). The very top of the mountain of sheltered areas may be permanently blanketed by snow even at the equator. Note that day length depends on latitude and the position of the sun—not altitude. Thus, seasonal variation in daylength does not contribute to these changes.

6. **Photoperiod:** Longer days promote increased community diversity and a more complex structure (though organism abundance and types may be restricted by other climatic variables, such as water availability). Very long days found in upper latitudes enable many organisms to complete their life cycle rapidly while conditions are favorable.
    Exposure: North-facing slopes have cooler microclimates and less insolation (the sun strikes

Exposure: North-facing slopes have cooler microclimates and less insolation (the sun strikes them at a more oblique angle). Therefore, community diversity may be lower on a north-facing slope and the organisms found there would be adapted to these environmental conditions.

Location in relation to the mountains: The climate on the lee side of mountain ranges is drier than on the windward side because, as air masses are forced up the windward side, they cool and loose their moisture. As they descend, they are dry and warm adiabatically ("Chinook" winds). Thus the community structure on either side of a mountain range may be quite different, with desert or grassland-adapted organisms on the lee side and moist (or even rain) forests on the windward side (e.g.,. eastern and western Oregon).

Wind velocity: Excessive wind velocity (as encountered at the tops of high mountains) will decrease community diversity and favor persistence of only those organisms adapted to intense wind stress and the drying conditions which accompany it. Note that blowing moisture and ice and snow can be quite abrasive and many plants may show "flagging" (branches are longer and healthier on one side of the tree, pointing away from the direction of prevailing winds).

Partial pressure of gases along an attitudinal gradient: The partial pressure of atmospheric gases decreases with altitude and the metabolic functioning of humans is impaired. (See the BioBytes program *Alien*, included on the CD-ROM.) As partial pressure of atmospheric gases decreases with altitude, the diversity of a community will decrease, and will eventually be dominated by the few organisms capable of adapting to these conditions. At higher altitudes, humans produce more red blood cells and those raised at very high altitudes will have increased pulmonary volume as well.

Permanently frozen substraum (permafrost): In regions of permafrost, such as the arctic tundra, soil thaws to only a few centimenters depth in summer and then refreezes. The subsoil is permanently frozen. Thus trees cannot grow, since their roots cannot penetrate the permafrost, and roots of herbaceous plants are disrupted by the freezing and thawing. Communities in permafrost regions are limited to small plants which can withstand the harsh environmental conditions and animals adapted to the long winters without abundant food supplies or that migrate to and from the tundra seasonally. Even though there is little precipitation in the tundra, melted water cannot percolate into the frozen subsoil so the summer habitat is characterized by boggy conditions in many areas. With long or continuous days, many insects with aquatic larvae develop rapidly and their multitudes form a food base for other insects and many shorebirds.

Ultraviolet (W) radiation: Excessive UV radiation limits the diversity of a community, restricting it to tolerant organisms. Evidence suggesting physical damage by UV radiation to plankton, frogs, and a variety of other organisms raises concern about the depletion of ozone and increased UV radiation due to human activity.

Ionizing radiation (cosmic rays, etc.), both natural and man-made: All life is continuously exposed to background radiation that comes from our rocks and soil and from cosmic rays. These radiation sources may be responsible for producing some of the genetic mutations that recur in all populations. Locally, certain communities may be exposed to higher levels of radiation and cosmic rays effects may increase with altitude. However, these imperceptible radiations have little, if any, impact on community structure. In the presence of man-made ionizing radiation (fall-out, X-rays, contamination with radioactive wastes), community diversity will be noticeably affected only at relatively high levels of radiation, with organisms restricted to the very few which are able to tolerate the conditions. Lower levels of radioactivity may also increase genetic diversity by inducing mutations and, because most mutations are harmful, may have serious affects on the future of restricted species exposed.

Soil type and texture: Various characteristics of soil, including particle size, porosity, drainage and chemical characteristics will affect the types of organisms. For example, the mostly organic, nutrient-rich topsoil of a mature forest supports a varied community of invertebrates and microorganisms, whereas less fertile soils, such as in deserts, support only limited communities. High salinity, acidity, presence of metallic ions, etc., may limit the community to tolerant species (e.g., plants and animals living on the dunes of ocean beaches are highly adapted to the presence of salt—they are "halophytes").

Laboratory 44

Predator-Prey Relations

EXERCISE A	Predation

p. 44-2 **a.** The predator has to forage farther afield. As the total number of prey are reduced, searching becomes more intensive.

b. The predator may systematically explore the area and return to explore areas previously searched.

p. 44-3 **Table 44A-1 Time Needed to Find Five Prey Items**

Predator	Time (seconds) for Each Day											
	1	2	3	4	5	6	7	8	9	10	11	12
Jerod	7	9	13	9	10	7	8	8	12	12	13	9
Azie	14	11	10	5	13	10	9	9	11	16	22	50
Keisha	8	13	14	6	23	17	55	32	5	D	D	D
Christi	6	10	20	45	55	30	20	10	40	40	D	D
Jennie	5	10	26	16	29	34	10	35	51	52	4	D
Kelly	15	11	9	13	10	17	17	38	25	55	D	D
Kelly J.	15	8	16	22	11	58	12	47	12	38	D	D
Kelly R.	14	7	17	24	27	16	37	28	13	20	15	D
Jennis	18	22	27	25	18	35	39	6	34	7	37	30
Lindsay	10	5	20	10	15	22	36	25	D	D	D	D
Kelly C.	8	7	9	10	5	20	10	10	20	50	D	D
Brit	9	7	9	8	8	7	15	12	13	28	50	D
Mary G.	12	11	12	20	30	20	14	7	34	D	D	D
JP	3	7	7	7	7	4	7	7	5	4	6	50
MEAN	10.3	9.9	14.9	15.7	18.6	21.2	20.6	19.6	23.9	35.9	40.5	52.8

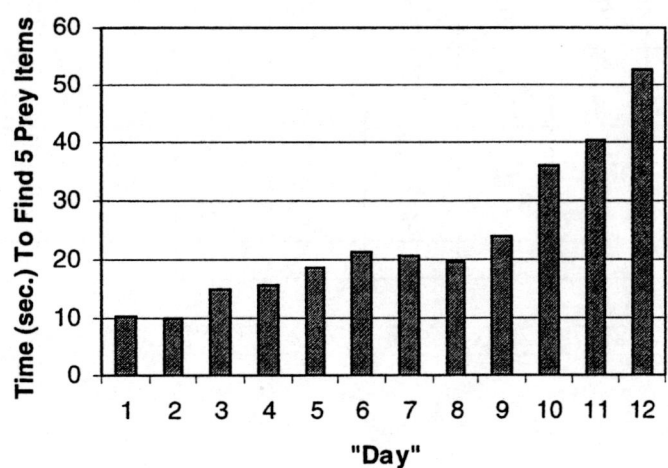

Note: Count "dead" predators as 60 seconds to obtain the Mean values used in this Exercise.

c. Any population foraging on a winter prey population (short-eared owls preying on meadow voles). The prey are non-replaceable since they are not reproducing so each capture reduces the pool from which further captures can be made.

d. Both predator and prey populations will decline.

e. If the prey are depleted beyond the level that will support the predator, predators may leave (disperse) to new areas in search of prey.

EXERCISE B Functional Responses by Predators

p. 44-4   a. The critical number of prey must be greater than 80 prey/12 days since no predators in Exercise B lived for the 12 days. In Exercise A, all predators were still alive on day 8 so 100 prey were enough to sustain all predators for that period. On day 9, a predator died. Therefore,

$$\frac{100}{8} = \frac{X}{12}$$

and $8X = 100/12$, $X = 150$, so the critical value would be 150 prey for 12 days.

b. The critical prey density would increase if the predator population increased, other predators used the same prey, temperatures were lower (if the predators are endotherms), or if additional predators immigrate.
The critical prey density would decrease if the predator population was reduced (by disease, age-related mortality, emigration, etc.) or warmer weather (if the predators are endotherms).

c. In Experiment B, the number of prey controls the predator population. The prey start at a certain number and do not reproduce. When they become scarce, the predators start to die because they cannot find prey items.

d. Yes, in Exercise A, as the prey items became scarce (after 8 days), predators began to die due to lack of food.

p. 44-5   **Table 44B-1 Time Needed to Find Five Prey Items at Various Prey Densities**

80 prey items

Predator	Time (seconds) for Each Day											
	1	2	3	4	5	6	7	8	9	10	11	12
Keisha	18	39	D	D	D	D	D	D	D	D	D	D
Kelly R.	15	17	40	9	10	24	23	40	D	D	D	D
Jennis	15	24	8	19	27	27	16	15	D	D	D	D
Brit	7	6	16	11	10	18	48	36	37	D	D	D
Mary G.	6	7	10	24	10	41	35	14	13	37	D	D
JP	4	4	8	9	26	23	21	D	D	D	D	D
MEAN	10.8	16.2	23.7	22.0	23.8	32.2	33.8	37.5	48.3	56.2	60.0	

60 Prey Items

Predator	Time (seconds) for Each Day											
	1	2	3	4	5	6	7	8	9	10	11	12
Christina	30	28	10	15	30	D	D	D	D	D	D	D
Jennie	5	5	7	10	12	20	45	49	35	10	D	D
Liza	14	14	11	8	4	24	48	D	D	D	D	D
Eliz	15	20	13	54	D	D	D	D	D	D	D	D
Kelly D.	10	33	30	20	20	15	D	D	D	D	D	D
Kelly J.	11	32	12	21	17	21	19	D	D	D	D	D
MEAN	14.2	22.0	13.8	21.3	23.8	33.3	48.6	58.2	55.8	51.7	60.0	

40 Prey Items

Predator	Time (seconds) for Each Day											
	1	2	3	4	5	6	7	8	9	10	11	12
Jerod	9	14	8	30	D	D	D	D	D	D	D	D
Azie	7	17	15	D	D	D	D	D	D	D	D	D
Lindsay	21	40	42	42	D	D	D	D	D	D	D	D
Kelly C.	35	55	D	D	D	D	D	D	D	D	D	D
MEAN	18.0	31.5	31.3	32	60.0							

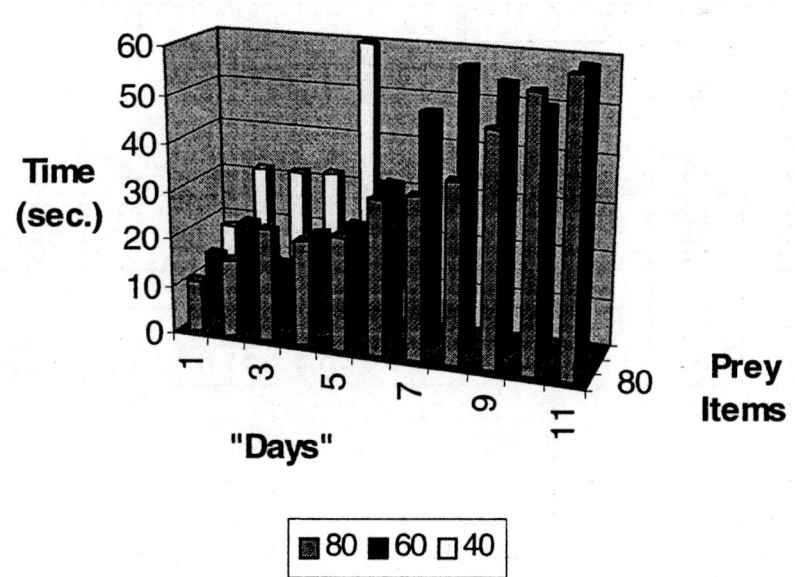

p. 44-6
   a. Exclusive use—depends on size and number of prey.
   b. Competitors could feed on the same prey, use the same resting or reproductive sites, disrupt the foraging behavior of the predator, or produce physical destruction of the habitat.
   c. Reproduction by prey would provide more food items for predators who would be declining if they did not reproduce so success should increase. If both reproduced, there would be a dynamic balance if there was sufficient time. Predator-prey oscillations in the far north have been documented and suggest that as prey increase (lemmings, snowshoe hares), predators (jaegers, (birds), Arctic foxes, etc.) also increase but lag somewhat. When prey reaches a peak and declines, predator mortality responds more slowly but results in fewer predators. This produces a continuous predator-prey "cycle" or oscillation over a number of years.

p. 44-7   Table 44C-1 Time Needed to Find Five Prey Items When Competitors Are Present

| Predator | Time (seconds) for Each Day | | | | | | | | | | | |
	1	2	3	4	5	6	7	8	9	10	11	12
Jerod	10	11	12	9	12	10	13	40	40	55	25	10
Azie	7	5	10	19	5	8	4	5	21	4	20	12
Keisha	16	8	4	7	7	10	22	15	43	10	10	11
Chris	10	5	5	8	12	4	8	8	10	17	15	31
Jennie	5	3	10	10	15	17	42	55	51	D	D	D
Liza	5	9	6	9	7	10	11	7	6	13	5	10
Eliz P.	11	12	33	10	18	10	7	37	24	14	D	D
Kelly	17	6	12	6	21	32	15	53	21	D	D	D
Kelly J.	15	17	7	32	43	27	55	D	D	D	D	D
Kelly R.	19	23	14	47	35	D	D	D	D	D	D	D
Jennis	14	6	9	29	19	48	28	D	D	D	D	D
Kelly C.	5	7	10	14	11	8	18	50	D	D	D	D
Brit	9	13	25	25	27	37	20	35	D	D	D	D
Mary G.	6	13	17	7	21	15	27	30	D	D	D	D
JP	10	12	5	4	4	9	16	3	6	8	4	5
MEAN	10.8	10.0	11.9	15.7	17.1	20.3	21.6	32.4	36.4	37.6	38.7	38.7

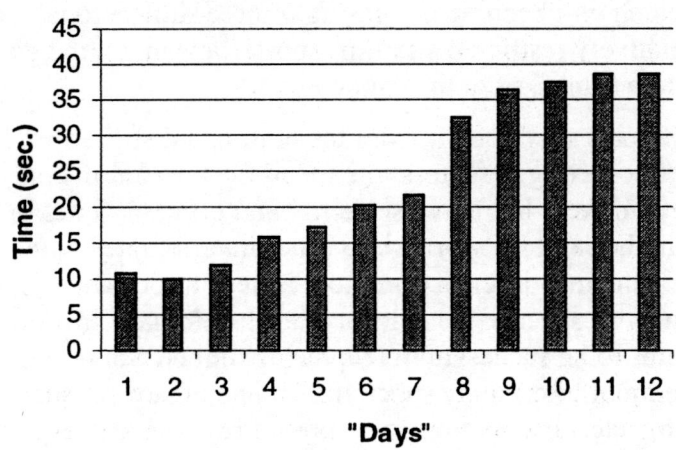

Laboratory Review Questions and Problems

1. Predators are usually larger than prey and are free-living, depending on hunting and chance encounter to locate prey. Parasites are smaller than their hosts and live either on the outside (ectoparasites) or inside (endoparasites) of their hosts. They both utilize the prey/host for energy/food but differ in their size and hunting strategies.

2. Predators and herbivores are both free-living and depend on hunting to locate their food reserves which are often patchily distributed. Animal prey have higher protein content and are more easily digested—but, being motile and capable of learning, can escape through both active and passive means (crypsis, etc.). Plants, on the other hand, are rooted to one spot and cannot move to escape. However, their food reserves are often harder to digest (cellulose in the cell walls cannot be digested directly by animals) and may evolve defenses including toxins (glycosides and other materials) that protect them from predators.

3. The following simulations could be used to explore a variety of situations:
   a. Crypsis—dye macaroni to match the background (green if the substrate is grassy).
   b. Toxic prey with warning coloration—dye macaroni red (warning) and penalize foragers 5 points for each item captured.
   c. Mimicry—dye macaroni red but place a piece of green rice inside (any forager that captures one of these items is *not* penalized (however, if the number of mimics is small relative to the models, the toxic prey, they will probably find this a maladaptive strategy).
   d. Alternative prey—supply other prey items (twists, marshmallows, etc.). What is the affect of size., abundance, and appearance of the alternative prey? (If large, abundant, attractive, and palatable, wouldn't the original prey benefit?)
   e. Use larger prey items (nuts).
   f. If prey are clumped, a predator may be better able to wipe out those that are close together (a family, village, etc.) or stay close and benefit from the association (prairie dog towns, etc.). If the prey are randomly distributed, the predator might benefit from a random search pattern. If they are evenly distributed, a regular, systematic search pattern might be helpful.

g. The greater the habitat complexity (layers, etc.), the harder it will be to locate prey because the search will become increasingly three-dimensional. However, prey often are located in a relatively restricted stratum (sub-terranean, on the ground, in low vegetation, in bushes, on tree trunks, or in the canopy).

4. (a) Predators would probably benefit from the clumped distribution, but many prey show a random dispersion, requiring predators to expend the maximum energy in their searches. (b) Blind predators are unlikely but may use tactile and olfactory cues to locate prey (many shorebirds probe in the sand for worms and other invertebrates without being able to see them). (c) Learning the responses of predators is very important to successful predation. Many young begin with specific behavioral patterns but have to learn the characteristics of their prey/food items to be successful in capturing dinner. Some, like oystercatchers (birds that feed on shelled mollusks), may specialize in one of several successful strategies (hammering, prying, etc.) for opening their prey. Prey also are responsive to the presence and activities of predators. Mobbing, in which various (often young) birds surround potential predators and display with loud noises, is a behavior that may reinforce the image and behavior of a predator.

5. Fitness: contribution of genes to the next gene pool.
   Coevolution: adaptation of species to each other (plants may benefit from specific insects or birds that act as pollinators so the length of the corolla tube and proboscis evolve synchronously to exclude competitors; plants evolve toxins to specific herbivores as the herbivores evolve resistance to the toxins; etc. Val Valen has proposed, in his "Red Queen hypothesis" that coevolution is a race in which there are no winners, only a continual battle to adapt.
   Niche: the role of an organism—what it does and how it lives. Its habitat is its address, but its niche is everything it does or is adapted to. G. Evelyn Hutchinson has proposed a multidimensional niche in hyperspace in which every property of the organism can be delimited. Organisms have a fundamental niche (determined by their genetics) and a realized niche (determined by their environment, including competitiors).
   Interspecific competition: competition between species. This is often most intense among more closely related species because, with a common heritage, they often have similar environmental requirements. Where closely related species come back into contact (alter allopatric speciation), they often compete and undergo character displacement to further separate their niche utilization and reduce competition (Darwin's finches, North American wood warblers, etc.). Where closely related species remain allopatric, they may retain very similar appearances (sibling species) but occupy very different niches and have different songs, behavior, etc. (e.g., *Empidonax* flycatchers in North America).
   Intraspecific competition: competition within a species. This is the most intense form of competition because all members of the species inhabit the same niche (unless males and females show sexual dimorphism and partition the niche to some extent based on size, feeding preferences, etc.)

Laboratory 45

Productivity in an Aquatic Ecosystem

EXERCISE A   Measuring Dissolved Oxygen: Effects of Temperature and Salinity

p. 45-3   Table 45-A1

Temperature	Titrant Used (ml)	DO (mg O_2/l)	Percent O_2 Saturation
20° C	0.89	8.9	95%

HYPOTHESIS: Cold water can hold more oxygen than warm water.

NULL HYPOTHESIS: The amount of oxygen dissolved in water is the same at all temperatures.

Prediction—as temperature increases, the amount of dissolved oxygen (mg O_2/l) will decrease.

Independent variable—temperature.

Dependent variable—amount of dissolved oxygen.

p. 45-4   n. No I_2 is left in the solution when the blue color disappears.

p. 45-6   Table 45A-2

Temperature	DO (mg O_2/l)	Percent O_2 Saturation
0° C	9.8	74
20° C	8.9	95
40° C	7.8	>100

Results support the hypothesis

Results allow the null hypothesis to be rejected.

The prediction was correct

Conclusion—as water warms, less oxygen is dissolved in the water. The capacity of water to hold oxygen also decreases. For this reason, water at 5° C represents only 74% saturation (because cold water has a greater capacity to hold O_2) while water at 40° C is more than 100% saturated (its capacity to hold O_2 is a lot less, so anything that helps introduce O_2 such as wind or currents can supersaturate the water if you consider its capacity to hold O_2).

EXERCISE B   MEASURING PRODUCTION IN AN AQUATIC ECOSYSTEM

p. 45-6   a. Net deposits are the useful amount (of energy) available for use.

p. 45-7   b. The first bar (PS) represents gross primary productivity.

c. The third bar (PS + R) represents net primary productivity. If the community also includes heterotrophs, the same bar also represents net community productivity.

p. 45-8   d. Yes. If there is any energy fixed at all it must be a positive value (negative uses are confined to respiration, export, etc.).

e. Yes. If metabolic utilization ("respiration") by autotrophs (or the community) does not exceed the energy fixed (gross primary productivity), the difference (net

---

productivity) is positive but if respiration exceeds the energy fixed, net productivity is negative (stored biomass is used). If respiration and gross productivity are equal, net productivity is zero—there is no accumulation or loss of biomass within the population or community.

p. 45-9 $NCP = -1.1$ mg O_2/l/hr
$R = 3.2$ mg O_2/l/hr
$GPP = 2.1$ mg O_2/l/hr

f. Oxygen declined because respiratory losses exceeded the amount of oxygen produced by photosynthesis.

g. Net community productivity should increase if photosynthesis speeds up (with new nutrients) but will slow when sunlight is reduced by clouds and overcast. If more heterotrophic bacteria are added they will increase respiration and this will decrease net community productivity.

p. 45-10   HYPOTHESIS: Respiration O_2 levels and gross productivity would be lower in an oligotrophic lake than in a eutrophic lake.

NULL HYPOTHESIS: There is no difference in the amount of respiration, O_2 level, and productivity in oligotrophic, eutrophic, and polluted lakes.

Prediction—Higher gross productivity will be found in eutrophic lake conditions.
Independent variable—type of lake.
Dependent variable—amount of dissolved oxygen.

p. 45-11   **Table 45B-2 Dissolved Oxygen Data for Sample**

Bottle	DO (mg O_2/l) (oligotrophic)
Initial	8.2
Dark	7.8
Light	9.0

Start time: ___8 AM___

Finish time: ___8 PM___

Duration ___12___ hr

p. 45-12   **Table 45B-3 Calculation of NCP and GPP**

Initial DO (mg O_2/l) ___8.2___

Dark DO (mg O_2/l) ___7.8___

Respiration rate (mg O_2/l/day) ___0.8___

Sample	(Light DO) DO (mg O_2/l) ppm	Net Community Productivity NCP (light – dark) (mg O_2/l/day)	Gross Primary Productivity GPP (light – dark) mg O_2/l/day
Oligotrophic	9.0	16	24

Table 45B-4 Class Data for __oligotrophic__ Sample

Group	NCP	GPP
1	1.6	2.4
2	1.2	1.6
3	1.5	2.1
4	0.8	1.7
5	1.8	2.2
Mean	1.38	2.0

Table 45B-5 Class Averages

Lake type	NCP	GPP
Oligotrophic	1.4	2.0
Eutrophic	1.6	2.8
Polluted	0	0

Results support the hypothesis but not the null hypothesis.

The prediction was correct.

Conclusion—gross productivity, net productivity, and O_2 levels are lower in eutrophic lakes. This is because there are fewer organisms to produce oxygen. Since algal cultures were used, we did not have to worry about respiration by heterotrophs.

p. 45-13 **h.** Oligotrophic—D

Eutrophic—A, B (more heterotrophs and slightly polluted)

Polluted—C

EXERCISE C   Productivity of Phytoplankton in a Water Column

p. 45-13 HYPOTHESIS: The more light present in the water column, the greater the amount of net and gross primary productivity.

NULL HYPOTHESIS: Productivity throughout a phytoplanktom column will not vary.

Prediction—bottles with more screens will show less gross and net productivity.

Independent variable—NPP and GPP.

Dependent variable—amount of light (screens).

p. 45-15 **Table 45C-1 Calculation of NCP and GPP**

Initial DO mg O_2/l ___5.4___		Start time: _2:30 PM_	
Dark DO mg O_2/l ___3.2___		Finish time: _2:30 PM_	
Respiration rate mg O_2/l/day ___2.2___		Duration: ___24___ hr	
Sample	**DO (mg O_2/l)**	**NPP or NCP (light – initial) (mg O_2/l/day)**	**GPP (light – dark) (mg O_2/l)**
Light	8.0	2.6	4.8
Light + N	8.8	3.4	5.6
Light + P	9.0	3.6	5.8

Table 45C-2 Class Averages for Productivity Data

Light	NPP (mg O$_2$/l/day)			GPP (mg O$_2$/l/day)		
	Light	Light + N	Light + P	Light	Light + N	Light + P
100%	2.6	3.4	3.6	4.8	5.6	5.8
65%	2.6	3.1	3.5	4.8	5.3	5.7
25%	2.0	2.6	2.8	4.2	4.8	5.0
10%	1.0	1.4	2.0	3.2	3.6	4.6
2%	−1.1	−0.7	−0.8	1.1	1.5	1.4

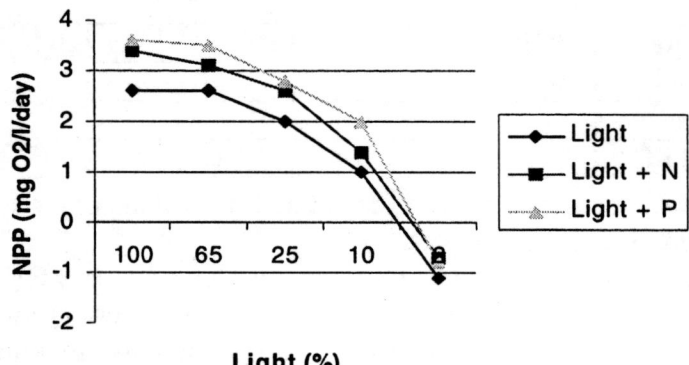

a. The samples in 2% light were light-limited. Productivity dropped in all samples at light intensities below 65%.

b. Phosphorus enrichment increased primary productivity, thus indicating some limitation by the initial phosphorus concentration of the water.

p. 45-16   Nitrogen enrichment also increased primary productivity, thus these water samples were also nitrogen limited

Results support the hypothesis but not the null hypothesis.

Conclusion—higher NPP and GPP occur in parts of a phytoplankton column that is exposed to more light.

Light bottle net primary productivity = ___975___ mg C/m^3/day

Light bottle gross primary productivity = ___1900___ mg C/m^3/day

Table 45C-3 Class Data on Productivity

Percentage of Incident Light Available	Net Primary Productivity (mg C/m^3/day)	Gross Primary Productivity (mg C/m^3/day)
100%	975	1800
65%	975	1800
25%	750	1575
10%	375	1200
2%	−413	413

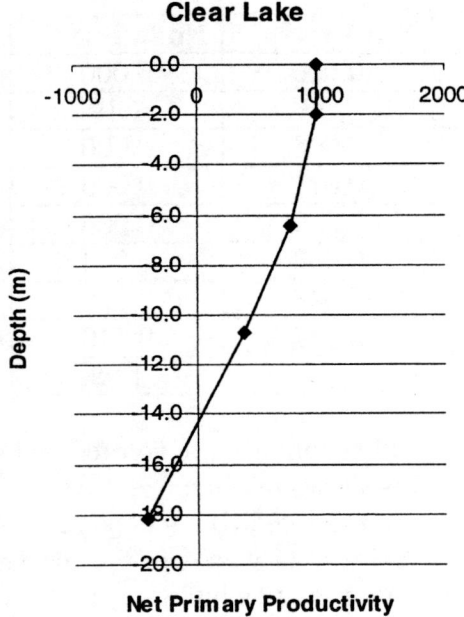

Clear Lake

Depth (m)

Net Primary Productivity

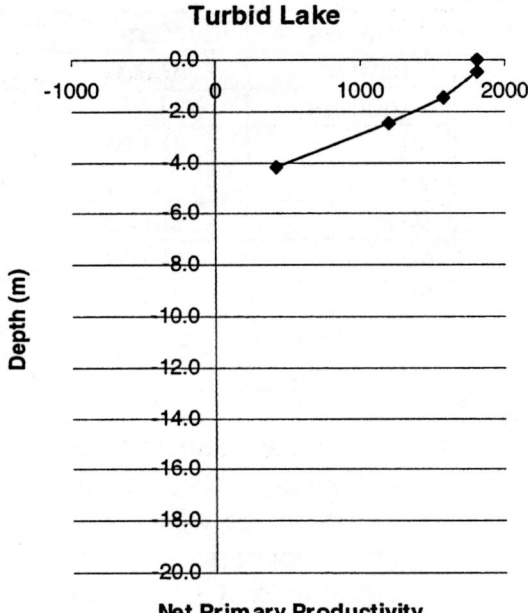

Turbid Lake

Depth (m)

Net Primary Productivity

p. 45-17 **c.** The surface waters receive effective sunlight. Light penetration is reduced with depth so photosynthesis and production decline (if the lake is deep enough, no light reaches the bottom and there is no photosynthesis).

 d. The compensation depth is about 3.4 m.

 e. The compensation depth would be nearer the surface on a cloudy day because there would be less light entering the water column but respiration of the organisms in the lake would remain the same (temperature would change little from day to day).

EXERCISE D   Thermal Stratification and Dissolved Oxygen Patterns in Lakes

p. 45-19 **a.** Thermal stratification exists because the epilimnion and the hypolimnion have different densities because they have different temperatures. In this experiment, the two layers differ in density because of different salt content. In both cases, stratification is based on a density difference.

 b. The "wind" is an energy source that causes circulation in the epilimnion. This water movement picks up small parcels of salty, colored water from the hypolimnion and disperses them. Without the wind, mixing could only occur because of molecular diffusion and very slow thermal convection currents.

 c. The absorbance of the surface solution will increase as more of the colored hypolimnion is mixed into it. The faster the wind speed, the faster the mixing.

 d. The absorbance difference between epilimnion and hypolimnion will decrease over time. If the two layers ever become completely mixed, there will be no difference at all.

 e. You would look for a lake with a sharp temperature (and therefore density) difference between surface and bottom waters. The greater the density difference between two layers, the more work the wind must do to mix them.

Time (min)	Fan 0 m	Fan 0.5 m	Fan 1 m	No fan
0 (surface)	0.000	0.000	0.000	0.000
0 (bottom)	1.20	1.18	1.23	1.19
10	0.140	0.048	0.008	0.000
20	0.132	0.064	0.011	0.000
30	0.163	0.048	0.050	0.000
40	0.170	0.058	0.058	0.004
50	0.174	0.060	0.048	0.010
60 (surface)	0.180	0.068	0.048	0.010
60 (bottom)	1.05			1.05

f. Yes, in the instructor's sample data, the top layer becomes more colored and he bottom layer becomes slightly less colored. This shows mixing.

g. The faster the wind speed, the more vigorous the mixing. Mixing depends on an energy input from the fan. On the other hand, the surface waters of the "no-fan" container only show a slight increase in absorbance over an hour.

h. No, even at the fastest wind speeds, the bottom waters remain much more colored than the surface waters.

i. A more concentrated hypolimnion would have made the "lake" more resistant to mixing. The absorbance of the surface waters would have increased more slowly.

j. A hot, sunny, calm day produces a warm epilimnion. The density difference between it and the hypolimnion would be great, and the weak wind would not be able to mix the two layers. A chilly, windy, cloudy day would reduce the temperature of the epilimnion, the density difference between it and the hypolimnion, wold be small, and the strong wind would probably be able to overcome this density difference and mix the two layers.

k. During the summer, thermal stratification is strong and the hypolimnion may become oxygen-depleted because it is shut off from the air. During the spring and fall, the whole lake is mixing, and all parts of the water column can absorb oxygen from the air. Also, during the spring and fall, temperatures are cooler, meaning that respiration is slowed down and oxygen solubility is increased.

Laboratory Review Questions and Problems

1. **a.** Neither the oxygen in the light bottle nor the oxygen in the dark bottle would change very much from the initial.
   b. The dark bottle's oxygen would decrease markedly, but the light bottle's oxygen would not change very much from he initial because, in the light bottle, respiration and photosynthesis are equal.
   c. The light bottle's oxygen would increase, but the dark bottle's oxygen would not change very much from the initial oxygen.
   d. The dark bottle's oxygen would decrease markedly, and the light bottle's oxygen would decrease almost as much

2. (a) 52% (b) 59% (c) 76%

3. The two oxygen profiles reflect the different patterns of light penetration and vertical distribution of phytoplankton in eutrophic versus oligotrophic systems. In the eutrophic system, oxygen concentration is maximal at the surface (often supersaturated), but is rapidly decreased in subsurface water due to light attenuation (by dense phytoplankton populations), which limits photosynthesis, and the relatively greater utilization of oxygen for catabolic processes in the subsurface water. In contrast, the depth distribution of oxygen in the oligotrophic lake is relatively invariant, due to enhanced light penetration through the clear (low population density) water. The region of change of oxygen concentration in this curve is probably at the thermocline (a region of rapid temperature change). Temperatures in the epilimnion (above the thermocline) are greater than in the hyplimnion (below the thermocline). The difference in oxygen solubility in warmer versus cooler water determines the shape of this curve, in addition to the potential displacement of the most photosynthetically active (or most dense) phytoplankton population to the deeper water.

4. The lack of oxygen in the hypolimnion of a eutrophic lake is caused by light limitation of photosynthesis due to absorption of light by epilimnetic phytoplankton. In an oligotrophic lake, phytoplankton population density is low throughout the water column, therefore epilimnetic cells do not shade hypolimnetic populations. Phytoplankton in the hypolimnion of an oligotrophic lake receive sufficient light for photosynthesis, therefore the hypolimnion is oxygenated.

5. The hypolimnion is cold and dark due to reduced penetration of light. It is calm because there is no (wind) turbulence in the deeper water. The nutrient-rich condition is a consequence of the degradation of senescent organisms settling through the hypolimnion from the epilimnion and dissolution of nutrients from the benthic sediments.

6. Phytoplankton in the surface waters of a eutrophic lake in summer would be limited by nutrient availability. Active photosynthesis and growth by a dense phytoplankton population will rapidly use up available dissolved nutrients (including the dissolved CO_2 required for photosynthetic "dark" reactions).

7. Curve A is produced by a dense phytoplankton population having maximal photosynthesis at the surface and rapid light limitation of photosynthesis beneath the surface, as might occur on a cloudy day when light intensity is low. Curve B could be produced by an inhibition of photosynthesis of a phytoplankton population at the lake surface, due to high light intensity. The maximum photosynthetic capacity (and often the greatest population density) is displaced to a subsurface depth where light intensity is lower. This type of curve is characteristic of many blue-green and other algae which have buoyancy-regulating mechanisms to adjust their position in the water column.

8. Thermal stratification occurs when sunlight warms the surface waters and these waters become warmer and less dense than the colder layers beneath them. The stability of the stratification (the amount of energy required to mix the layers) becomes greater as surface water temperature increases.

   The epilimnion has warmer water, higher dissolved oxygen, enhanced primary productivity, and is often deficient in one or more limiting nutrients. The hypolimnion has cooler, nutrient-rich water, less dense phytoplankton populations and light is often limiting.

Notes:

Alien: A Simulation of Cardiopulmonary Physiology

Variables Displayed on the Digital Screen

p. A-5   **a.** The oxygen saturation of venous blood measures the oxygen remaining in blood after use by tissues and is therefore a measure of the metabolic oxygen demand.

           b. 0.28 mm Hg.

           c. Metabolism (especially during active exercise) generates CO_2 that enters the venous blood.

EXERCISE A   Demonstration Experiments

PART 1   Exercise at Sea Level

p. A-8   **a.** Exercising muscles consume oxygen. Hemoglobin readily releases oxygen to oxygen-starved tissues. There is less oxygen in the venous blood because relatively more oxygen was released to the tissues.

p. A-9   **Table AA-1 Human Response to Exercise at Sea Level (Oxygen, 159 mm Hg)**

	Resting	Exhaustion	0.5 Minutes After Collapse*	1.0 Minutes After Collapse*
Ventilation	6.4	107.77	67.42	27.19
Heartbeats/minute	70	204	160	110
Venous P_{O2}	40	23	18	18
ArterialP_{CO2}	40	46	41	39
Arterial pH	7.4	7.22	7.30	7.34

      * Time of collapse: 0.6 minutes

           b. Ventilation increases in response to increased concentrations of CO_2 in the blood.

           c. The faster blood flow to tissues is, the more oxygen the blood can bring in and the more CO_2 it can carry away.

           d. Oxygen in the venous blood drops only slightly, from 40 mm Hg to approximately 20-30 mm Hg.

           e. Arterial P_{CO2}. Ventilation is very sensitive to arterial P_{CO2}. Once production of CO_2 diminishes, ventilation can easily restore arterial P_{CO2} to resting levels.

           f. Ventilation and heart rate return to resting values as activity ceases. As activity ends, production of CO_2 from the tissues decreases, therefore arterial P_{CO2} is less and arterial pH increases due to lessened CO_2 and lactic acid inputs.

PART 2   Exercise at High Altitude

p. A-10   **a.** Lack of oxygen caused the failure of homeostasis. Increased ventilation and heart rate cannot deliver sufficient oxygen to the tissues when the atmospheric concentration of oxygen is less than one-third normal (normal atmospheric oxygen concentration at sea level is approximately 159 mm Hg).

   b. Venous P_{O2} diminishes rapidly, reaching the "lethal level" of 5 mm Hg, after which the subject dies. In the sea level exercise experiment, collapse was followed by a return to homeostasis.

   c. Ventilation and removal ("blowing off") of CO_2 is relatively easy in a high altitude, "thin" atmosphere. After cardiac arrest and cessation of ventilation, CO_2 can no longer be removed from the tissues due to lack of blood flow. Therefore, arterial P_{CO2} increases rapidly, reaching 57 mm Hg just before brain death.

p. A-11   **Table AA-2 Human Response to Exercise at 9,000 m (Oxygen, 50 mm Hg)**

			Venous P_{O2}	
	Resting	**Unconscious**	**5 mm Hg**	**0 mm Hg**
Ventilation	6.41	118.4	116.9	0
Heartbeats/minute	70	200	199	0
Venous P_{O2}	40	10	5	0
Arterial P_{CO2}	40	39	38	39
Arterial pH	7.43	7.34	7.32	7.30

PART 3   **Increasing Atmospheric CO_2**

p. A-12   **Table AA-3 Human Response to Increasing Partial Pressures of CO_2**

	0 mm Hg	**30 mm Hg**	**60 mm Hg**	**120 mm Hg**
Ventilation	6.51	21.5	109.04	0
Arterial P_{CO2}	40	43	62	120
Arterial pH	7.43	7.40	7.24	6.96

   a. Arterial P_{CO2} and pH remain relatively constant. Arterial P_{CO2} only increases by 3 mm Hg despite an increase of 30 mm Hg in the atmosphere.

   b. Ventilation rate increases to approximately 20 liters/minute in order to clear CO_2 from the tissues.

   c. Yes. The maximum rate of ventilation is approximately 109 liters/minute. Ventilation at this rate cannot compensate for the additional atmospheric CO_2 at 60 mm Hg.

   d. Ventilation rate cannot increase beyond 109 liters/minute. Therefore, arterial CO_2 increases as atmospheric CO_2 approaches 120 mm Hg. Under these conditions, the respiratory center in the brain is anesthetized. Therefore, ventilation decreases as the subject collapses.

EXERCISE B   **The Effect of Altitude on Running Stamina**

p. A-14   **Table AB-2 Effect of Decreasing Partial Pressure of Oxygen on Human Running Performance**

	$P_{O2}(159)$	$P_{O2}(120)$	$P_{O2}(90)$	$P_{O2}(75)$	$P_{O2}(70)$
Venous P_{O2}	18	17	13	9	9
Arterial P_{CO2}	42	41	41	40	40
Arterial pH	7.27	7.27	7.28	7.33	7.38
Distance run	465	400	339	210	105

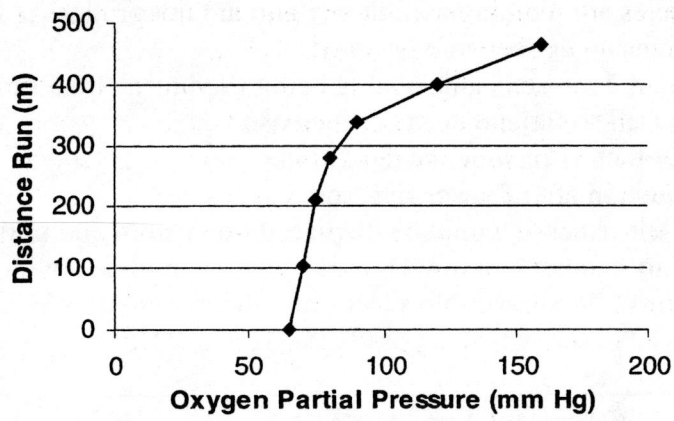

Figure AB-2 *Graph of the relationship between partial pressure of oxygen and running stamina*

p. A-15 **a.** This few minutes acclimation to environmental conditions assures that oxygen remaining in the hemoglobin is appropriate to the new environmental conditions.
b. Unconsciousness occurs without exercise at about $P_{O_2} = 67$ mm Hg.
c. Increased ventilation and heart rate remain insufficient to supply adequate oxygen to the tissues. Therefore, the subject becomes unconscious.
d. No. Stamina would still be limited by the supply of muscle glycogen and the ability of the body to remove wastes such as CO_2 and lactic acid from the tissues.

EXERCISE C   Running Distance and Running Speed

p. A-16 **Table AC-1 Distance Run before Exhaustion at Various Running Speeds**

	10 m/sec	8 m/sec	6 m/sec	4 m/sec	3 m/sec
Venous P_{O_2}	28	24	20	18	20
Arterial P_{CO_2}	48	46	44	40	39
Arterial pH	7.21	7.22	7.26	7.29	7.30
Distance run	228	290	380	630	1650

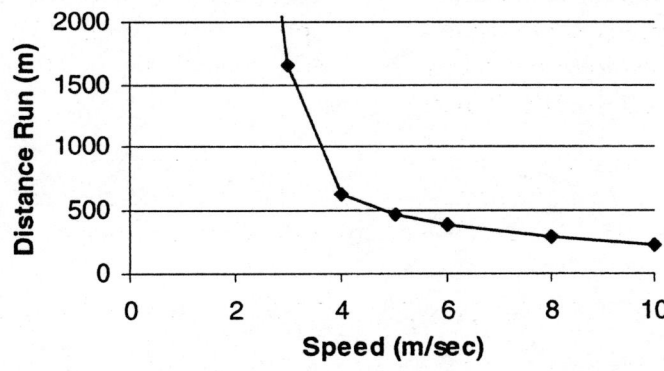

Figure AC-1 *Graph of the relationship between running speed and distance run.*

p. A-17　**a.** Muscles are working aerobically and are not producing lactic acid. Therefore the muscles do not become fatigued.

　　　b. Yes. At 2 m/sec., lactic acid is being produced at the same rate it is being degraded so fatigue does not increase.

　　　c. Other factors include the depletion of muscle glycogen, which would cause exhaustion after four or five hours.

　　　d. The entire curve would be displaced downward and to the left. Therefore, a human running in a low O_2 atmosphere would be able to run only shorter distances at comparable speeds and the maximum speed would be 8 m/sec (rather than 10 m/sec which is possible at 159 mm Hg).

EXERCISE D　　　Answering the Questions

p. A-18　**Table AD1 Answers to the 16 Questions in Exercise D**

Question	Human	Vega	Aurigae	Ceti 2	Ceti 3	Regulus
1. Resting oxygen consumption rate (l/min)	0.19	1.39	0.95	0.02	0.286	4.15
2. Exercising oxygen consumption rate (l/min)	1.07	3.33	1.80	0.033	0.367	4.85
3. Resting heart rate (beats/min)	69	170	36	12	23	110
4. Exercising heart rate (beats/min)	153	451	51	17	27	147
5. Resting mean arterial pressure (mm Hg)	98	131	524	12	89	238
6. Exercising mean arterial pressure (mm Hg)	132	208	540	16	108	255
7. Total blood volume (liters)	5.6	7.1	12.5	2.18	16.7	12.8
8. Resting cardiac output (l/min)	5.15	3.2	22.46	0.22	31.05	6.65
9. Exercising cardiac output (l/min)	15.36	17.56	32.54	0.97	66.15	8.41
10. Resting cardiac stroke volume (l/min)	74.6	18.82	624	18.3	1350	60.4
11. Exercising cardiac stroke volume (l/min)	100	38.9	638	5.71	2450	57.2
12. Exercising cardiac output increased by faster beat or by larger stroke volume	beat	beat	beat	stroke	stroke	beat
13. Lung volume (liters)	5.55	1.29	32	10.0	9.2	0.303
14. Maximum speed at which a resting subject can run (m/sec)	10	16	8	4	2	5
15. Running distance (m) at 50% of maximum rate	465	1056	456	108	310	233
16. Ability to live in earth's atmosphere	yes	no	no	yes	yes	no

Cycle: A Simulation of the Menstrual Cycle and Human Fertility

EXERCISE A   The *Cycle* Game

p. C-9

a. A woman is fertile for only a brief period during her menstrual cycle because the egg remains viable for only the first 24 hours after it is released and sperm cannot last longer than four days in the female reproductive tract. Therefore, the only time "window" in which intercourse (with ejaculation) can cause pregnancy is from four days before ovulation to one day after it.

b. Either estrogen levels or follicle diameter are best for predicting the time of ovulation because they must reach a characteristic level before ovulation occurs.

c. Anything that prolongs the survival of egg and sperm widens the fertile period. If sperm could survive in the female reproductive tract for one month instead of four days, one intercourse per month, at any time, would be sufficient to guarantee fertilization.

d. The greatest increase in basal body temperature occurs after ovulation, and thus does not predict the onset of ovulation. By the time that basal body temperature starts to rise, a woman is in her fertile period.

e. A woman's basal body temperature is rising right after ovulation, so this is a fertile period. However, once this rise starts, the egg is losing fertility fast.

f. After ovulation, the follicle is converted to the corpus luteum. If the ovum is not fertilized, the corpus luteum degenerates within approximately fourteen days, and menstrual flow begins. Thus, menstrual flow begins fourteen days after ovulation. The time interval between menstrual flow and the next ovulation is less constant.

EXERCISE B   The Effect of Birth Control Pills

p. C-10

a. High levels of both progesterone and estrogen levels decrease secretion of LH and FSH by the pituitary.

b. Progesterone inhibits LH production and thus prevents ovulation. Estrogen inhibits FSH secretion so that follicles do not develop. Since constant high estrogen levels cause the endometrium to thicken (in preparation for implantation of an embryo), there is no menstrual flow until estrogen levels are decreased.

c. Very low levels of estrogen and progesterone increase FSH secretion so the follicle grows rapidly. However, it can't mature. Also, without sufficient estrogen production to act as a trigger, there is no LH surge. Therefore, the follicle never ovulates and becomes atretic. The uterine lining also does not thicken sufficiently to maintain an implanted zygote.

Notes:

Dueling Alleles: A Simulation of Genetic Drift and Selection

Overview

p. D-1 **a.** Evolution is a change in allele frequencies. If allelic frequencies remain constant (under the conditions listed above), evolution will not occur.

EXERCISE A   Genetic Drift

p. D-3 **Table DA-1 Number of Generations to Extinction of Fixation of an Allele in A Population (No Selection)**

Pairs	2 Mating Pairs	4 Mating Pairs	20 Mating Pairs	50 Mating Pairs
b extinct	13			541
b fixed		54	47	71
Mean time to extinction	13.2	28.1	137.2	429.6

c. Both alleles persisted longer in the populations with the larger numbers of mating pairs.

d. Smaller. Fluctuations will be larger with a smaller population because in smaller populations, random mating outcomes can have a larger effect.

EXERCISE B   Selection

p. D-5 **Table DB-1 The Effect of Recessiveness on the Persistence (in Generations) of a Deleterious Allele**

	Generations to Extinction of *b* Allele			
	B does not mask *b*		*B* masks *b*	
bb death probability	Your Results	Class Results	Your Results	Class Results
1.0	5	5.6	11	18.7
0.5	14	14.0	25	40.5
0.2	49	41.3	65	74.7

p. D-6 **a.** The deleterious trait is not expressed if the recessive deleterious allele occurs in a heterozygous individual. Therefore the deleterious allele only is selected against when it is homozygous.

b. Yes.

c. Both recessive and nonrecessive deleterious alleles will persist longer in the population when the selection pressure against them is lower from the start (e.g., death probabilities of 1.0, 0.5, and 0.2).

d. The frequency of the *b* allele declines most rapidly when the *b* allele is abundant.

The proportion of homozygous recessive individuals is greater in a population in which the b allele is more abundant.

   e. Hinder.

   f. Recessiveness will hinder the spread of a beneficial allele because natural selection can only operate upon individuals displaying the beneficial phenotype, and only homozygous recessive individuals will display the beneficial phenotype.

p. D-7   3. There should be fluctuation without an upward or a downward trend.

   5. The trend should be declining.

p. D-8   **a.** Heterozygote advantage is the condition in which heterozygotes have greater reproductive success than either type of homozygote. The preferential selection of the heterozygote has the effect of sheltering recessive alleles which may be lethal in homozygotes.

   b. Once the heterozygote advantage is removed, the frequency of the recessive allele would diminish due to random genetic drift and the lack of reproduction by homozygous recessive individuals.

   c. Yes. Unbalanced selection can maintain both alleles for long periods. However, the frequency of b will be lower than its frequency in the balanced selection case, and drift might cause b to go extinct sooner than it would in the case of balance selection.

Seedling: A Computer Simulation of Plant Growth and Plant Competition

Overview

p. S-1 **a.** The tertiary structure of the protein is altered as the bonds (such as H-bonds) which maintain tertiary structure are broken.

b. Denatured enzymes are structurally incapable of complexing with their specific substrates, and thus cannot catalyze chemical reactions. The uncatalyzed rate of most reactions is too slow to support metabolic processes in plants.

c. Proteins and nucleic acids.

EXERCISE A   Spacing Plants for Optimal Yield

PART 1

p. S-3 **a.** Excessive crowding, such that several layers of leaves overlap, will result in a leaf area index that is much greater than 1. The major consequences of this overlap is a reduction in light available for photosynthesis for the shaded leaves. Plants will tend to become spindly as they grow, with relatively fewer leaves. Additional consequences of crowding include intense competition for water and nutrients.

p. S-4   **Table 27SA-1 Plant Performance under Moist Conditions**

	Interplant Distances					
	1 cm	3 cm	6 cm	9 cm	12 cm	15 cm
Ratio of leaf area to soil area	12	1.33	0.33	0.14	0.08	0.05
Soil water depletion	50.08	5.56	1.39	0.61	0.34	0.22
Percent of water demand satisfied	100	100	100	100	100	100
Percent of maximum net photosynthesis rate	0	61	82	85	86	86
Net photosynthesis/plant (mg C/day)	−1.2	7.22	10.09	10.5	10.63	10.69
Number of plants/m^2	10,000	1,111	278	123	69	44
Net (areal) photosynthesis (mg C/m^2/day)	−11,992	8,024	2805	1296	738	475

Figure SA-1(a) *Photosynthesis per plant (smallest plants)*

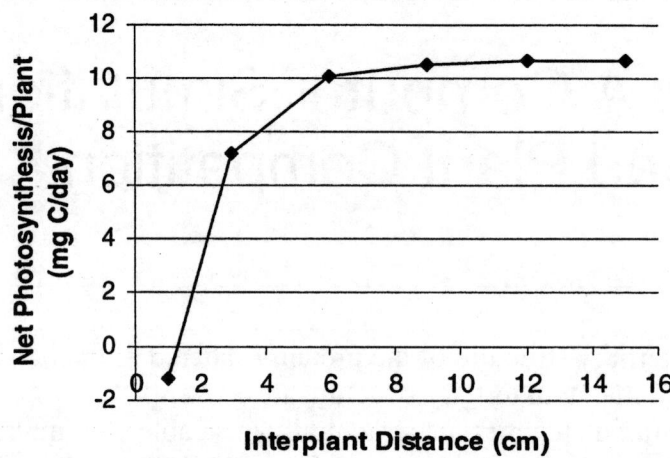

Figure SA-1(b) *Photosynthesis per square meter (smallest plants)*

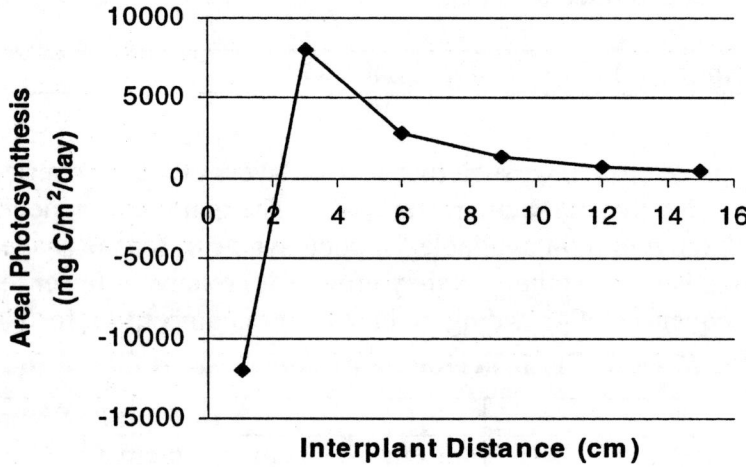

b. At very close interplant distances, plants compete for available light, water and nutrients.

c. Leaf area index decreases as interplant distance increases. Plants that are further apart are less likely to shade each other. Soil water depletion increases as interplant distance decreases, since water demand is greater for denser plantings.

d. For the moist environment, a leaf area index of about 8 will permit positive net photosynthesis. The dry environment cannot maintain positive net photosynthesis with a leaf area index any greater than 2. Thus, the "breakeven" leaf area index is not a constant, but varies with conditions.

e. Net photosynthesis for individual plants increases as interplant distance increases. The rate of increase is greatest in the transition from very small to moderate Interplant distances (e.g., 1 to 3 cm).

f. No. The further apart are plants, the more rapidly each of them will photosynthesize. There is no interplant distance at which *individual* net photosynthesis is maximized. The main effect is "dilution" of photosynthesis of individual plants by larger and larger interplant distances.

g. The optimum interplant distance varies with the size of the plant and the environment. For the smallest plants, it is 3 cm in the moist environment and 4 cm in the dry environment, for the middle-sized plants it is 4 cm (moist) and 6 cm (dry), and for the largest plants it is 5 cm (moist) and 8 cm (dry). Thus the optimum interplant distance will increase as the plants grow.

PART 2

p. 27-6   **Table SA-2 Plant Performance Under Dry Conditions** (small-sized plants)

	Interplant distances					
	1 cm	3 cm	6 cm	9 cm	12 cm	15 cm
Ratio of leaf area to soil area	12	1.33	0.33	0.14	0.08	0.05
Soil water depletion	14.72	4.73	1.44	0.66	0.38	0.24
Percent of water demand satisfied	6	18	22	23	24	24
Percent of maximum net photosynthesis rate	0	11	18	20	20	21
Net photosynthesis (mg C/plant/day)	−1.2	0.38	1.39	1.59	1.67	1.7
Number of plants/m^2	10,000	1,111	278	123	69	44
Net photosynthesis (mg C/m^2/day)	−12,000	432	387	197	116	75

- **a.** Evidence that these conditions are stressing the plants includes: (1) the diminished net photosynthetic rate/plant; (2) percent of water demand satisfied; and (3) percent of maximum net photosynthetic rate under dry conditions.
- **b.** Closing stomata reduces net photosynthesis, because gas exchange through the leaves is restricted. Gas exchange is necessary to provide CO_2 to leaf cells for fixation in the Calvin-Benson cycle and to diminish potentially inhibitory high oxygen concentrations in the leaf.
- **c.** The net photosynthetic yield/plant is consistently approximately six times greater under moist conditions (at all interplant distances other than 1 cm).
- **d.** Yes. No. Prolonged opening of the stomata in a dry environment would result in desiccation of the leaves.
- **e.** Areal photosynthesis is consistently lower in the dry environment (at all interplant distances greater than 1 cm). These differences occur because water is required for photosynthesis and growth (e.g., turgor pressure is significant in cell elongation).

EXERCISE B   Temperature, Sunlight, and New Photosynthesis

p. S-8   **Table SB-1 Net Photosynthesis at Various Temperatures and Light Intensities** (medium-sized plants)

	Light intensities (% full sunlight)				
Temperatures	5%	15%	30%	50%	90%
3° C	1	2.7	3.5	4.4	4.5
10° C	1.9	4.7	6.1	6.6	7.4
15° C	2.3	6.7	9.3	8.6	10.4
22° C	4.8	10.9	14.8	17.1	16.3
33° C	−1.6	4	6.7	9.7	10.7

p. S-9   **a.** Rate of photosynthesis increases with light intensity because light is the required energy source for photosynthesis. When light is dim, a doubling of light intensity can lead to a doubling of photosynthetic rate. However, at very high light intensities, the increase in photosynthetic rate is not as great, for the limiting factor becomes, not the quantity of light available, but the metabolic rate of light-independent photosynthetic reactions.

b. At very low temperatures, the plant metabolic rate and the functioning of photosynthetic enzymes is reduced. Increasing the quantity of light available does not increase photosynthetic rate when the overall metabolic rate is limited.

Figure SB-1 *Net photosynthesis at different temperatures and light intensities (medium-sized plants)*

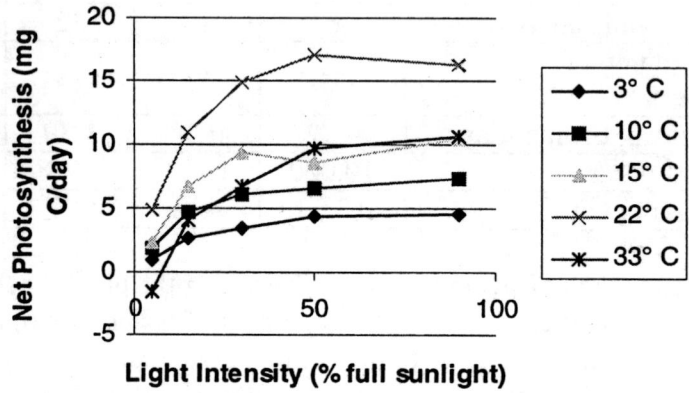

c. The temperature optimum for photosynthesis is approximately 25°C. Above this temperature, the process of photorespiration consumes photosynthate as fast as it is created, therefore net photosynthetic rate (measured here as the amount of fixed carbon) decreases.

d. The compensation point is lower at lower temperatures.

p. S-10   **e.** The compensation point varies with temperature because at high temperatures respiration is more rapid, and a faster rate of photosynthesis (requiring a higher light intensity) is necessary to balance respiration. In addition, Figure 1 in the *Preparator's Guide* for Seedling shows that the optimum temperature for photosynthesis is 25°C, while the optimum temperature for respiration is 35°C. Thus, as the temperature rises over 25°C, photosynthesis is declining while respiration is increasing. Thus a plant at temperatures over 25°C requires more light to make up for respiratory losses than a plant which is cooler than 25°C.

f. Hot, dry conditions would be especially damaging to plant growth because, at the same time as the plant's respiration is high, its photosynthesis would be reduced by closed stomata and lack of carbon dioxide.

p. S-11   **Table SC-1 Soil Water Depletion at Various Temperatures and Humidities**
(medium-sized plants)

	Leaf Temperature (° C)				
Relative humidity	5°	10°	15°	20°	25°
5%	2.3	3.2	4.1	5.7	8.6
15%	1.3	1.8	2.5	3.6	4.5
30%	0.9	1.0	1.7	2.4	2.8
60%	0.4	0.5	0.6	1.0	1.4
85%	0.1	0.1	0.2	0.2	0.4

p. S-12   **Figure SC-1** *Soil water depletion (medium-sized plants)*

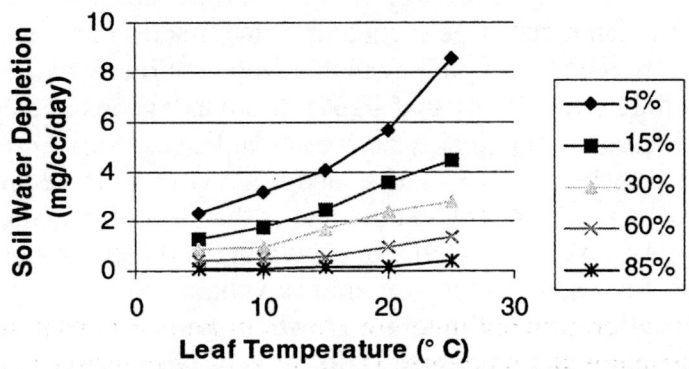

a. Evaporation from leaves increases with temperature.

b. Transpiration rate decreases as the water potential difference between the atmosphere and the inside of the leaf is lessened.

c. On very hot, dry days, a plant may restrict water loss by closing its stomata.

d. Amounts of water lost would be lower and would not increase as much as humidity fell because the plant would be limited by the amount of water in the soil.

e. Transpiration is an unavoidable side effect of open stomata, and open stomata are necessary for carbon dioxide uptake.

f. Desert plants are structurally and metabolically adapted to cope with hot, dry conditions. Plants typically have very thick cuticles (to minimize evaporation), small or modified leaves (e.g., cactus, in which the leaves have been reduced to spines and the stems are photosynthetic), and life cycles that condense vegetative growth, flowering and seed production into the brief periods of time when adequate water is available. In addition, many desert plants exhibit a modified photosynthetic process (CAM) whereby the stomata open only at night (when transpiration is lowest) and store CO_2 in organic acids until dawn. During the day when light is available for photosynthesis (and stomata are closed), CO_2 is released from the stored organic acids and used in the synthetic (dark) reactions of photosynthesis.

EXERCISE D   The *Seedling* Game

p. S-16   **a.** Resource allocation is the channeling of available food/energy supplies to various parts of an organism. Here the resource is photosynthate variously allocated to augment plant height, stem diameter, leaves or root area.

p. S-17   **b.** Early in a plant's life cycle, photosynthate is usually allocated to support vegetative growth. As growth continues, photosynthate produced by the leaves supports development of flowers and fruits. Allocation strategies also change in response to environmental pressures. For example, rapid initial growth of new leaves expands the photosynthetic capability of the plant and leaf maturation reduces herbivory.

   c. Yes. Animals that hibernate during cold or otherwise adverse environmental conditions often accumulate considerable reserves of body fat during the time of the year immediately prior to the start of the hibernation.

   d. In desert plants, the large succulent stems function in water storage, the large root systems maximally exploit available water sources and the tiny leaves reduce transpiration. Increasing leaf size is disadvantageous in a dry environment, for transpiration would quickly desiccate the leaves. Rapid growth rate is unnecessary in a desert where interplant distance is likely to be large and there is no competition for sunlight. (Though note that desert annuals, which complete entire life cycles of vegetative growth, flowering and fruit production within the few months when moisture is available, are characterized by rapid growth rates).

   e. An allocation strategy favoring growth of large diameter stems and extensive root system, and which minimized allocation of photosynthate to the leaves (which should be reduced in size) would be successful in semiarid Texas. In Puerto Rico, allocation of photosynthate to the production of large leaves and tall stems would enhance the plant's ability to compete for light in the tropical, rainy (reduced sunlight) environment.

   f. See the *Seedling* section of the *Preparator's Guide* for a discussion of the "Causes of Death in Seedling." The plants least likely to die are short plants having thick stems and few leaves.